VIRUS
LIFE *in*
DIAGRAMS

Hans-W. Ackermann, M.D.
Laurent Berthiaume, Ph.D.
Michel Tremblay, Ph.D.

VIRUS LIFE *in* DIAGRAMS

CRC Press

Boca Raton Boston London New York Washington, D.C.

Cover diagram after H.R. Gelderblom, M. Gentile, A. Scheidler, M. Özel, and G. Pauli, Zur Struktur und Funktion bei HIV: Gesichertes, neue Felder und offene Fragen, *AIDS Forsch.*, 8, 231, 1993. Courtesy of R.S. Schulz-Verlag, Starnberg-Percha, Germany, and Springer-Verlag, Vienna. With permission.

Library of Congress Cataloging-in-Publication Data

Catalog information may be obtained from the Library of Congress

CONTENTS

ACKNOWLEDGMENTS

This book was produced with the collaboration of several major scientific publishers. Academic Press provided 64 diagrams originating from its New York or London divisions. This large number reflects the outstanding role of Academic Press as the publisher of *Advances in Virus Research, Virology,* and *Journal of Molecular Biology.* The American Society for Microbiology, publisher of the *Journal of Virology* and many important books, provided 30 diagrams. Other major sources of diagrams were Springer-Verlag, Heidelberg and Vienna (25); Raven Press, New York (23); Jones and Bartlett, Boston (16); CRC Press (14); and the Amsterdam, Cambridge, and Paris divisions of Elsevier (10).

We wish to thank the following persons for permission to use their figures in this book: U. Aebi, D. Anderson, D.C. Ansardi, E. Arnold, E.J. Arts, D. Baltimore, D.H. Bamford, K.I. Berns, K. Bienz, D.H.L. Bishop, L.W. Black, G.W. Blissard, A.D. Boothe, M. Botchum, R.F. Boyd, S.M. Brookes, P.O. Brown, K.W. Buck, P.J.G. Butler, A.J. Cann, S. Casjens, J.L. Casey, E. Catalano, B. Caughey, E.C.S. Chan, G. Cohen, R.W. Compans, R.A. Consigli, D.H. Coombs, S.N. Covey, R.L. Crowell, C. Danglot, J. Das, G. Dehò, G. Devauchelle, N.J. Dimmock, B. Dreiseikelmann, R. Dulbecco, A.K. Dunker, M.K. Estes, T.W. Fenger, F. Fenner, H. Furukawa, H.R. Gelderblom, R. Geleziunas, J.-C. Georges, A.A. Gershon, R. Goldbach, G. Gosztonyi, W.C. Greene, R. Grimaud, J.C. d'Halluin, W. Hammerschmidt, C.U.T. Hellen, R.W. Hendrix, T. Hohn, F.L. Homa, J.H. Hoofnagle, R.W. Horne, E. Jawetz, W.K. Joklik, C. Kedinger, M. Kielian, A. Kornberg, E. Kutter, G.M. Lenoir, T.L. Lentz, B.H. Lindqvist, B.E.L. Lockhart, J. Maniloff, M. Marsh, D.A. Marvin, W. Mason, C.K. Mathews, J.L. Melnick, S. Michelson, L. Mindich, P. Model, G. Mosig, B. Moss, K.J. Murti, M. Nassal, M.N. Oxman, P. Palese, J.L. Patterson, R.F. Pettersson, B. Plachter, B.V.V. Prasad, C.M. Rice, D. Riesner, B. Roizman, R.R. Rueckert, M. Salas, M.S. Salvato, C.E. Samuel, C.S. Schmaljohn, F.L. Schuster, P. Serwer, A. Severini, L.D. Simon, K. Simons, L.S. Sturman, R.H. Symons, J.M. Taylor, G.J. Thomas, A.J. Tingle, G. Toolan, P. Traktman, J.K. VanSlyke, L.E. Volkman, R.R. Wagner, R.E. Webster, R.B. Wickner, P.R. Wills, and W.H. Wunner.

E. Catalano, B. Caughey, K. Dunker, H.R. Gelderblom, E. Kutter, and S. Michelson provided updated or improved versions of already published diagrams. Jeannine Gauthier, Quebec, and Zahira Chouchane, now in Setif, Algeria, carried out bibliographical research. Suzanne Bernatchez and Lyna Pelletier, Laval University, provided editorial help in the final stages of the preparation of the manuscript. Martin Berthiaume, Ville-de-Laval, recorded or reconstituted numerous diagrams. Dr. Morris Goldner and Philippe Cantin, Laval University, gave helpful advice on presentation.

THE AUTHORS

Hans-Wolfgang Ackermann, M.D., is Professor of Microbiology at the Medical School of Laval University, Quebec, Canada. He was born in Berlin, Germany, in 1936 and obtained his medical degree in 1962 at the Free University of Berlin (West). He was a fellow of the Airlift Memorial Foundation and received much of his microbiological training at the Pasteur Institute of Paris.

After a period of research and teaching at the Free University, he left Germany in 1967 and went to Canada, where he started to investigate bacteriophage morphology. During his career, he has done research on enterobacteria, airborne fungi, hepatitis B virus, and baculoviruses; however, bacteriophages have always been at the center of his interest. He teaches virology and mycology and has a strong interest in audiovisual teaching aids. In 1982, he founded the Félix d'Hérelle Reference Center for Bacterial Viruses, which is essentially a culture collection aimed at the preservation of type phages. Dr. Ackermann is author or co-author of about 140 scientific papers or book chapters and senior author of two books, entitled *Viruses of Prokaryotes* and *Atlas of Virus Diagrams* (CRC Press, 1987 and 1995, respectively). He has been a member of the International Committee on Taxonomy of Viruses (ICTV) since 1971 and several times was chairman or vice-chairman of its Bacterial Virus Subcommittee, was ICTV vice-president from 1984 to 1990, and was a member of the ICTV Executive Committee until 1996. He is now a life member of the ICTV.

Laurent Berthiaume, Ph.D., is a former professor at the Virology Center of the Armand-Frappier Institute, a branch of the University of Quebec at Montreal. He was born in Montreal in 1941 and in 1972 obtained his Ph.D. degree in microbiology–immunology at the University of Montreal. He was trained in electron microscopy at the University of Toronto School of Hygiene, was in charge for many years of the electron microscopic laboratory of this institute, and eventually became coordinator of graduate studies and registrar.

During his activities as a virologist and electron microscopist, he developed a strong interest in virus morphology and replication and the rapid diagnosis of viral infections. For many years, he lectured on virus structure and taxonomy. His most recent work has centered on viruses infecting fish, especially on the antigenic and genetic diversity of aquabirnaviruses. He is co-author of the *Atlas of Virus Diagrams* and the author of about 85 scientific papers or book chapters and 150 communications at congresses. He has been an ICTV member since 1975 and was a member of the ICTV Executive Committee from 1989 to 1995. He is presently retired and a free-lance writer in virology.

Michel Tremblay, Ph.D., is Professor of Microbiology at the Medical School of Laval University. He was born in Jonquière, P.Q., in 1955 and in 1989 obtained his Ph.D. degree in experimental medicine from McGill University. After post-doctoral work in molecular immunology at the Montreal Clinical Research Institute, in 1991 he founded a laboratory for human retrovirology at the Laval University Hospital. Dr. Tremblay is the author or co-author of 50 papers and 85 communications at congresses or symposia.

His research centers on immunological and virological aspects of the human immunodeficiency virus (HIV), the etiologic agent of AIDS. He is studying signal transduction after contact between the virus particle and its target cell, the modulatory role of intracellular tyrosine phosphatases on regulatory sequences of HIV, and the acquisition of host-derived glycoproteins by nascent HIV and their role in the viral life cycle. Dr. Tremblay is a member of the International AIDS Society.

Chapter 1

INTRODUCTION

This book is a sequel to the *Atlas of Virus Diagrams,* published in 1995 and concerned with viral morphology.[1] Its aim is to assemble the many diagrams of viral life cycles, particle assembly, and strategies of nucleic acid replication that are scattered over the literature. As was its predecessor, this book is directed at teachers and students. The title reflects the fact that viruses are inert in the environment and come to life during their multiplication within living cells.

Virology makes extensive use of diagrams to illustrate particle morphology and morphogenesis, viral genetic and replication cycles, epidemiology and replication of viral diseases, and virological techniques. Their function is to integrate large numbers of data of different origin, from electron microscopy to amino acid composition and clinical observations, into a single picture or a sequence of related diagrams. Photographs and experimental data usually illustrate the state or activity of a virus at a given moment. Diagrams transcend this moment and are thus ideally suited to illustrate the time course of viral life cycles and the steps of viral nucleic acid replication. Some diagrams, intended to promote understanding and future research, contain speculative elements and some reflect personal views or interpretational errors of the day. Others are historical documents and time capsules illustrating the development of virology. For example, the nature of lysogeny is explained in a diagram from 1953, and a series of drawings from 1962, featuring virus-infected cells, shows that the cytopathic effects and general life cycles of many mammalian viruses were known at an early date. Finally, many diagrammatical representations of viral development are true artistic creations that deserve preservation in their own right.

The 233 diagrams in this book were selected from about 1000 drawings from over 50 periodicals and 150 books or monographs from the English, French, and German literature. Most diagrams were from a few core journals, notably *Advances in Virus Research, Virology, Journal of Virology,* and *Journal of General Virology.* Scientific content was the prime criterion for selection, followed by clarity and didactic value, originality of concept, esthetic appeal, and historic interest. Vertebrate viruses and the tailed bacteriophages T4 and λ were well illustrated in the literature, often with considerable detail. Representations of retrovirus and herpesvirus life cycles were particularly frequent, but many other, generally less known viruses were not illustrated at all. As a consequence, the diagrams in this book illustrate the morphogenesis and replication of only 30 virus families and six nonclassified

"floating" genera. Plant viruses are clearly underrepresented and as many as 23 families and 15 "floating" genera of bacterial, plant, fungal, algal, or protozoal viruses are absent from this book. In plant viruses, this reflects the centering of research interests on applied rather than molecular microbiology. In fungal, algal, and protozoal viruses, this reflects lack of funding.

The diagrams are arranged according to the nature of viral nucleic acids and replication strategies rather than virus morphology or host range. The various types of viral nucleic acids, single-stranded or double-stranded RNA, have evolved many different strategies for replication, generally in relationship with their use of host mRNA. Consequently, the diagrams are grouped into five chapters corresponding to (i) a series of comparative diagrams, (ii) DNA viruses, (iii) viruses using reverse transcription, (iv) RNA viruses, and (v) miscellaneous entities including prions.

The comparative diagrams illustrate the properties of large groups such as enveloped or ssRNA viruses. A diagram on interferons is added because of its general nature. DNA viruses are subdivided into viruses with ssDNA (certain bacteriophages and parvoviruses), dsDNA viruses with cubic or helical symmetry, and dsDNA viruses with binary symmetry or tailed bacteriophages; this latter section is warranted because of the large number and diversity of diagrams available for these phages. Viruses using reverse transcription comprise DNA and RNA viruses of plants (badna- and caulimoviruses) and vertebrates (hepadna- and retroviruses). The chapter on RNA viruses includes sections on viruses with (+) sense and (−) sense ssRNA and viruses with dsDNA. The first two sections contain viruses with segmented and nonsegmented genomes. This arrangement brings out taxonomically important common properties. For example, regardless of morphology or host specificity, (+) strand RNA virus genomes code for RNA polymerases and replicate in strikingly similar ways, (−) sense RNA viruses are enveloped, (−) sense ssRNA and dsRNA viruses contain a viral RNA-dependent RNA polymerase, and dsRNA viruses have segmented genomes which all seem to be transcribed within viral particles. Similarly, one notes that several virus groups with dsDNA, namely adeno-, herpes-, and iridoviruses, resemble tailed phages in several aspects of replication and particle assembly.

Virus families and "floating" genera are arranged alphabetically within each section. Each family or single genus is characterized by a few key words and the introductory paragraph includes a short description of taxonomic status, particle morphology, host range, and major physiological features. In families illustrated more than once, diagrams are arranged from the general to the detailed, starting with cytopathic effects or whole life cycles and ending with the replication of nucleic acids. Except for retroviruses, there are few illustrations of genomic maps, translation, and regulation. The reasons are lack of space, due to the need to limit the number of pages of this book, and the uneven state of knowledge in virology, a few viruses being known in exquisite detail and the vast majority in outlines only. Virology has not yet reached the point where a coherent picture of translation can be given. Similarly, diagrams illustrating fine points of virology, for example the priming of DNA replication in tectiviruses or the biosynthesis of influenza virus hemagglutinin, have been left out because they were considered to be too detailed.

The diagrams were recorded with a Super Vista S-12 high-resolution scanner (UMAX Technologies, Freemont, CA; 600 × 1200 dpi). Many captions and some damaged diagrams, for example drawings recovered from microfilm, were restored and a number of foreign-language captions were translated into English. Legends represented particular problems. Some are extremely long in the original, extending over whole pages of small print. Others are so short that they are, taken out of context, not informative. We consider that each legend is an independent unit that has to be, in principle, understandable without consulting the original publication (although this may still be necessary in very complex diagrams). This situation required extensive rewriting. Many legends were shortened while still preserving their information content and others were completed from the accompanying text.

An overview of the state of virus classification, including the latest taxonomic changes, is presented in Chapter 2. Descriptions of individual taxa may be found in the periodical reports of the International Committee of Virology. An outline of viral multiplication, centering on mammalian viruses and regrettably short in information on viruses of fungi, algae, and protozoa, is given in Chapter 3. The diagrams are followed by an extensive glossary, warranted by a profusion of new terms or alternative meanings. Many frequently used terms, e.g., "capsid," "core," or "matrix," have acquired several meanings; even the now famous term "provirus" not only applies to the latent integrated DNA of retrovirus genomes, but is also used to designate immature virus particles. Many definitions in the glossary are taken or derived from the virological dictionary of Hull, Brown, and Payne[2] and the recent *Encyclopedia of Molecular Biology*.[3] Others have been devised by us.

Chapter 2

A SUMMARY OF VIRUS CLASSIFICATION

Viruses consist minimally of nucleic acid and a protein shell or capsid. The shell is of cubic or helical symmetry or, in tailed phages, a combination thereof ("binary symmetry"). Capsids with cubic symmetry are icosahedra or related bodies. In about one third of virus families, the capsid is surrounded by a lipid-containing envelope. A few exceptional types have an envelope and no capsid. The nucleic acid is single-stranded or double-stranded DNA or RNA, is linear or circular, and comprises one or several molecules (up to 12). All DNA viruses except polydnaviruses contain a single molecule of DNA. All RNA genomes except that of hepatitis deltavirus are linear. In some plant viruses, the individual segments of multipartite genomes are packaged into separate shells, constituting multicomponent systems.

The present edifice of virus classification is essentially the work of the International Committee on Taxonomy of Viruses or ICTV. It developed from the International Committee on Nomenclature of Viruses (ICNV), which was established in 1963, became permanent in 1966 at the XIth International Congress of Microbiology in Moscow, and was renamed ICTV in 1973.[4, 5] The ICTV consists of 6 subcommittees, 45 study groups, and over 400 participating virologists. It has subcommittees for viruses of vertebrates, invertebrates, plants, bacteria, eukaryotic protists (algae, fungi, protozoa), and the creation of a viral database. Study groups are formed by co-optation as the need arises. Taxonomic and nomenclature proposals usually originate in study groups and are successively voted on by the respective subcommittees, the ICTV Executive Committee, and the full ICTV when it convenes every three years at an International Congress of Virology. Only then do they become official.

The ICTV is concerned with a coherent, universal system of virus classification and nomenclature and issues periodically a report, in principle after each International Congress of Virology. The report describes virus orders, families, subfamilies, and genera. Species are simply listed without further details. The Sixth Report was published in 1995. Contrary to previous reports, it contains numerous micrographs, genomic maps, and diagrams illustrating replication strategies.[5]

In principle, viruses are classified with the help of all available data. Some criteria are of particular value, namely the presence of DNA or RNA, type of capsid symmetry, and the presence or absence of an envelope. These criteria were introduced by Lwoff, Horne, and Tournier[6] as the basis of a hierarchical system of viruses. The system did not survive, but its major criteria stood the test of time. More high-level criteria were added later. They relate to the genome (strandedness, message-sense [+] or anti-message [–] strands, number of segments) and replication (presence of a DNA step or RNA polymerase). Most criteria are virus-dependent. Host-related criteria such as symptoms of disease are few and of limited importance. Genome structure and base or amino acid alignments are becoming increasingly important as data become available. The ultimate goal is a phylogenetic system of viruses. Viruses are essentially classified by the following criteria:

1. Nucleic acid: DNA or RNA, number of strands, conformation (linear, circular, superhelical) and segments, sense (+ or –, ambisense), nucleotide sequence, G+C, presence or absence of 5′-terminal caps, terminal proteins, or poly(A) tracts.
2. Morphology: size, shape, presence or absence of an envelope or peplomers, capsid size and structure.
3. Physicochemical properties: particle mass, buoyant density, sedimentation velocity, and stability (pH, heat, solvents, detergents).
4. Proteins: content, number, size, and function of structural and nonstructural proteins, amino acid sequence, glycosylation.
5. Lipids and carbohydrates: content and nature.
6. Genome organization and replication: gene number and genomic map, characteristics of replication, transcription and translation, post-translational control, site of particle assembly, mode of release.
7. Antigenic properties.
8. Biological properties: host range, mode of transmission, geographic distribution, cell and tissue tropisms, pathogenicity and pathology.

The present system of viruses includes 2 orders, 53 families, 9 subfamilies, and over 160 genera. The order *Mononegavirales,* characterized by the presence of a single molecule of negative-sense ssRNA, comprises the *Filoviridae, Paramyxoviridae,* and *Rhabdoviridae* families. The recently created order *Nidovirales* includes the families *Arteriviridae* and *Coronaviridae*. Tailed bacteriophages are likely to become a third order. Families and genera are the most stable parts and the backbone of the system. About 20 "floating" genera, mostly of plant viruses, are not yet classified into families. The *Poxviridae* family has 11 genera, but other virus families are monogeneric and some even consist of a single member. The ICTV has also adopted the polythetic species concept, meaning that a species is defined by a set of properties, all of which are not necessarily present in all members. A virus species is defined as a "polythetic class of viruses that constitutes a replicating lineage and occupies a particular ecological niche."[7]

Names of virus taxa always reflect some virus characteristics. Order, family, subfamily, and genus names end with the Latin suffixes *-virales, -viridae, -virinae,* and *-virus,* respectively. The creation of Latinized binomial names for virus species is not attempted. Table 1 summarizes the present state of virus classification and includes the taxa approved at the Xth International Congress of Virology in Jerusalem (1997). Virus taxa are listed by the nature of nucleic acid, presence or absence of an envelope, and capsid symmetry. While most viruses have cubic or helical nucleocapsids, a few types with envelopes and no apparent capsids (*Fuselloviridae, Plasmaviridae,* the genera *Deltavirus* and *Umbravirus*) cannot be attributed to any symmetry class.

Table 1
VIRUS FAMILIES AND NONCLASSIFIED GENERA

Genome	Envelope	Capsid Symmetry	Order or Family	Floating Genera	Number of Genera	Host[a]
ssDNA	No	Cubic	*Circoviridae*		1	P
			Geminiviridae		3	P
			Microviridae		4	B
			Parvoviridae		6	V, I
	No	Helical	*Inoviridae*		2	B
dsDNA	Yes	Cubic	*Herpesviridae*		7	V
			—	ASFV[b]	1	V
	Yes	Helical	*Baculoviridae*		2	I
			Lipothrixviridae		1	B
			Polydnaviridae		2	I
			Poxviridae		11	V, I
	Yes	None	*Fuselloviridae*		1	B
			Plasmaviridae		1	B
	No	Cubic	*Adenoviridae*		2	V
			Corticoviridae		1	B
			Iridoviridae		4	V, I
			Papovaviridae		2	V
			Phycodnaviridae		1	A
			Tectiviridae		1	B
			—	*Rhizidiovirus*	1	P
	No	Binary	*Myoviridae*		6	B
			Siphoviridae		5	B
			Podoviridae		3	B
Viruses using reverse transcription						
dsDNA	Yes	Cubic	*Hepadnaviridae*		2	V
	No	Cubic	—	*Badnavirus*	1	P
			—	*Caulimovirus*	1	P
dsRNA	Yes	Helical	*Retroviridae*		7	V

Table 1 (continued)

Genome	Envelope	Capsid Symmetry	Order or Family	Floating Genera	Number of Genera	Host[a]
ssRNA, (+) sense	Yes	Cubic	*Flaviviridae*		3	V, I
			Togaviridae		2	V, I
			Nidovirales			
	Yes	Helical	*Arteriviridae*		1	V
			Coronaviridae		2	V
	Yes?	?		*Umbravirus*	1	P
	No	Cubic	*Astroviridae*		1	V
			Bromoviridae		4	P
			Caliciviridae		1	V
			Comoviridae		3	P
			Leviviridae		2	B
			Nodaviridae		2	V, I
			Picornaviridae		6	V, I?
			Sequiviridae		1	P
			Tetraviridae		2	I
			Tombusviridae		5	P
			—	*Enamovirus*	1	P
			—	*Ilaeovirus*	1	P
			—	*Luteovirus*	1	P
			—	*Necrovirus*	1	P
			—	*Sobemovirus*	1	P
			—	*Tymovirus*	1	P
	No	Helical rods	*Barnaviridae*		1	F
			—	*Furovirus*	1	P
			—	*Hordeivirus*	1	P
			—	*Tobamovirus*	1	P
			—	*Tobravirus*	1	P
	No	Helical filaments	*Closteroviridae*		1	P
			Potyviridae		3	P
			—	*Capillovirus*	1	P
			—	*Carlavirus*	1	P
			—	*Potexvirus*	1	P
			—	*Trichovirus*	1	P
			Mononegavirales			
(−) sense, monopartite	Yes	Helical	*Bornaviridae*		1	V
			Filoviridae		1	V
			Paramyxoviridae		5	V
			Rhabdoviridae		5	V, I, P
		?	—	*Deltavirus*	1	V
(−) sense, multipartite	Yes	Helical	*Arenaviridae*		1	V
			Bunyaviridae		5	V, I, P
			Orthomyxoviridae		4	V,
	No	Helical	—	*Tenuivirus*	1	P
dsRNA	Yes	Cubic	*Cystoviridae*		1	B
		?	*Hypoviridae*		1	F
	No	Cubic	*Birnaviridae*		3	V, I
			Partitiviridae		4	F, P
			Reoviridae		9	V, I, P
			Totiviridae		3	F, Pr

[a] A, algae; B, bacteria; F, fungi; I, invertebrates; P, plants; Pr, protozoa; V, vertebrates.
[b] ASFV, African swine fever virus.

Chapter 3

THE REPLICATION CYCLE

A striking feature of viruses is their specialized lifestyle as obligate intracellular parasites. No virus can survive and multiply without an appropriate host. Although many viruses carry replication-related enzymes, none are able to grow on their own and all have complex life cycles imposed upon them by their parasitic nature.

The biology of viruses was initially studied using bacterial viruses or bacteriophages as models. Two significant experiments with bacteriophages permitted the discovery of the fundamental nature of viruses. The "one-step growth" or "single-burst" experiment described by Ellis and Delbrück[8] in 1939 was the first to distinguish three essential phases of virus replication, namely the initiation of infection, replication and expression of the viral genomic material, and release of mature virus particles. "Single-burst" experiments using eukaryotic viruses revealed that the major difference between these viruses and bacteriophages was the generally much longer time required for replication, measured in hours and sometimes days rather than minutes. The slower growth rate of eukaryotic cells and the generally more complex nature of the replication cycle of eukaryotic cells are responsible for this difference. Hershey and Chase[9] carried out in 1952 a second key experiment using bacteriophage T2. Labeling proteins with ^{35}S and nucleic acids with ^{32}P, they showed that the DNA genome of the bacteriophage T2 entered the cell and started an infection, while the viral shell remained outside. This indicated that the genetic information was contained in the nucleic acid.

Virus infection of a host cell normally results in a full replication cycle with production of progeny viruses, but may also be abortive, restrictive, or latent. Abortive infection is due to expression of only a portion of the viral genome or to defective virus particles. The restrictive type of infection is seen in cells that are transiently permissive for virus replication or in cell types in which only a fraction of the cell population produces viral progeny at any time. Latent infection is characterized by the persistence of viral genomes within the cell and the expression of some viral genes, but without production of infectious viral progeny and cell destruction.

The viral replication cycle is arbitrarily divided into six partially overlapping stages, namely attachment, penetration, uncoating, replication, assembly, and release. All viruses go through these steps or their equivalents. Infection, penetration, uncoating, and release vary according to the unicellular or pluricellular nature of the host and the presence or absence of viral envelopes. The mode of replication varies with the nature of viral nucleic acids.

Attachment. Virus attachment results from the binding of a virion protein to a molecule located on the surface of the target cell (receptor) and is the first event in the viral infection of a cell. Viral binding sites are distributed over the entire viral surface or are located at specific places such as the tail tip of tailed phages. Cellular receptors are distributed over the whole cell or are restricted to specific areas, for example the basolateral surfaces of polarized epithelial cells or sex pili of bacteria. In animal viruses, the target receptor molecules are mostly glycoproteins.

Attachment is the first factor to influence the "tropism" of a virus for particular host species, organs, tissues, or cells. Blocks at later stages of replication occur in some cases and may modulate the virus host range. Attachment to a cellular receptor may affect the structure of the virion. Adsorption of phages to bacteria is also highly specific and involves the participation of specialized viral fixation organelles (e.g., tail fibers or spikes) and bacterial receptors, which may be located on the bacterial capsule, cell wall, flagella, pili, and even the plasma membrane. Plant viruses do not seem to have such attachment specificities.

Penetration. Viruses infect cells by entering them or by injecting their nucleic acid. Cell surface proteins without any role in virus attachment may have essential roles in viral entry. For example, the human immunodeficiency virus requires chemokine receptors (e.g., CXCR4, CCR5, CCR3, CCR2b) as cofactors to facilitate its entry into target cells. Similarly, vitronectin and mannose-6-phosphate are accessory factors in the entry of adenoviruses and herpesviruses, respectively. In a few cases, pinocytosis/phagocytosis leads to virus entry without interaction with specific receptors. This is seen with antibody-coated virus particles fixed to Fc receptors located on the surface of monocytes/macrophages. In this case, enhancement of virus infection is seen instead of the expected antibody-mediated virus neutralization.

Penetration, unlike attachment, is an energy-dependent step and takes place rapidly after attachment. Three major mechanisms are involved: (i) translocation of the complete viral particle across the cellular cytoplasmic membrane, (ii) endocytosis of the virion into intracellular vacuoles, and (iii) fusion of viral lipid envelopes and the cellular membrane. Nonenveloped viruses utilize the first two mechanisms, while enveloped viruses enter cells by fusion at neutral pH. The presence of a specific fusion protein in the virus envelope is needed for fusion (e.g., influenza virus hemagglutinin or retrovirus transmembrane glycoproteins). Endocytosis is the normal way eukaryotic cells take in materials bound to the cell surface. Endocytosis of viruses generally involves receptors and is mediated by clathrin-coated pits and vesicles. The newly formed endosomes fuse with other intracellular vesicles and release their contents into larger vesicles which are acidified by a cellular proton pump to pH 5.5–6.5. This causes conformational changes of virion coat proteins leading to fusion of viral and vesicle membranes. The nucleocapsid is then free to enter the cytoplasm of the infected host. Release of virus particles from endosomes and their passage into the cytoplasm are closely linked to the process of uncoating.

Because of the presence of barks, waxes, and cuticles and because of their thick, rigid cell walls, plants are not readily infected by contact and most of their viruses have to be introduced by mechanical wounding. Under natural conditions, this is generally done by biting or piercing arthopods (aphids, beetles, leafhoppers, mites, thrips) or nematodes, and even by parasitic fungi. Once the plasma membrane has been exposed, plant viruses may enter the cell by fusion with the plasma membrane (enveloped viruses) or simple endocytosis (nonenveloped viruses).[10] Viruses spread further by plasmodesmata and sap. In most bacteriophages, the nucleic acid enters the cell and the empty capsid remains on the outside. Only one phage, the enveloped cystovirus φ6, penetrates through the cell wall. Fungal viruses are generally transmitted horizontally and intracellularly during cell division, hyphal fusion, and spore production. The transmission of algal and protozoal viruses is poorly understood. Figure 226 suggests that some protozoal viruses have evolved special infective strategies.

Uncoating. The release of nucleocapsids from envelopes, their complete or partial disintegration, and the exposure of viral genomic material are referred to as uncoating. In animal viruses, uncoating of enveloped viruses takes place either at the cell membrane inside endosomes (pH-dependent membrane fusion with acidification as the triggering event) or directly in the cytoplasm (pH-independent). The fusion of viral envelopes and endosomal membranes is driven by the exposure, due to conformational changes induced by the low pH inside the vesicle, of a previously hidden virus fusion domain. Several compounds prevent acidification of endocytic vesicles (e.g., lysosomotropic agents such as ammonium chloride and chloroquine) and can be used to determine whether virus entry is mediated via pH-dependent or -independent membrane fusion. In bacteriophages, uncoating generally takes place at the cell surface and is accompanied by penetration of the nucleic acid. Tailed bacteriophages digest the bacterial cell wall locally with muralytic enzymes and inject their DNA through the plasma membrane. Filamentous phages virtually dissolve at the cell surface and only their DNA enters the cell. The coat proteins of many plant viruses with (+) ssRNA are removed by ribosomes attaching to the 5′ end of the nucleic acid during translation.[11]

Replication. The strategy of replication is directed by nature and conformation (DNA or RNA, single-stranded or double-stranded, linear, circular) of the viral genetic material and the pathway of mRNA synthesis. Generally, replication is semi-conservative and proceeds uni- or bidirectionally by strand displacement. Conservative RNA replication is found in reoviruses. There are several interesting generalities. For example, the majority of RNA viruses multiply in the cytoplasm, while all eukaryotic DNA viruses, excluding poxviruses, multiply and assemble in the nucleus. DNA viruses frequently replicate via a rolling-circle mechanism, which is also used by the circular RNAs of hepatitis D virus and viroids. On the other hand, some viruses have rare, particular features which result from or influence replication. For example, tailed bacteriophage DNAs may have circular permutations of sequence, terminal repeats, terminal proteins, cohesive ends, single-stranded gaps, unusual bases (e.g., 5-hydroxymethylcytosine or 5-hydroxymethyluracil), or DNA-bound sugars.

Viral nucleic acids can be divided into seven groups according to their replication strategy. This classification is based on the original six groups proposed by Baltimore[11] in 1971 and includes the replicative strategy of the so-called pararetroviruses.

Class I, dsDNA. This class can be subdivided into two groups depending on whether replication occurs exclusively within the nucleus (*Adenoviridae, Herpesviridae, Papovaviridae*) or the cytoplasm (*Poxviridae*). In the first case, replication depends heavily on cellular factors for transcription (e.g., cellular DNA-dependent RNA polymerase II), while the virus replicative cycle is independent of the cellular machinery in the second group. Because in many viruses different mRNAs come from different DNA strands, the designation of (+) and (−) strands is not meaningful.[11]

Class II, (+) ssDNA. Replication of parvoviruses (vertebrates and insects) occurs in the nucleus and involves the formation of a (−) sense strand, which serves as a template for the synthesis of a (+) strand. The infecting DNA of microviruses and inoviruses (bacteriophages) is made into dsDNA after synthesis of a complementary strand. This replicative form generates via a rolling-circle mechanism progeny replicative forms which, in turn, serve to synthesize novel viral DNA.

Class III, dsRNA. This class includes viruses with segmented genomes, notably the *Reoviridae* and *Cystoviridae*. Virions contain RNA-dependent RNA polymerase. Each segment is transcribed separately to produce (+) sense RNA molecules that will serve as both mRNAs and as templates for the synthesis of a complementary (−) RNA strand to form the mature double-stranded genome. In reoviruses, replication is conservative and parental RNA is retained in viral cores in the cytoplasm. In cystoviruses, replication is semi-conservative and viral RNA is synthetized by strand displacement.

Class IV, (+) ssRNA. The viral RNA acts as mRNA and is infectious. RNA polymerase is not present in the virion, but is coded for by the viral RNA. This class is subdivided into two groups, namely viruses with RNA which is translated in a single round to produce progeny protein (*Caliciviridae, Flaviviridae, Leviviridae, Picornaviridae*) and viruses with a complex transcription pathway necessitating two rounds of translation (*Nidovirales, Togaviridae*).

Class V, (–) ssRNA. Virions contain RNA-dependent RNA polymerase. The viral RNA is noninfectious and must be transcribed by the polymerase into (+) RNA for both transcription and translation. This class comprises viruses with segmented (*Arenaviridae, Bunyaviridae, Orthomyxoviridae*) and nonsegmented (*Mononegavirales*) genomes. In the first group, the initial step in replication is the production of monocistronic mRNA from each segment. Replication in the second group is similar except that multiple monocistronic mRNAs, for each of the virus genes, are produced from a nonsegmented genome.

Class VI, (+) ssRNA with a DNA intermediate in replication (Retroviridae). The virus genome is a (+) sense diploid RNA that is converted by a reverse transcriptase (RNA-dependent DNA polymerase) present in the virion into linear dsDNA (provirus) which is then integrated into the host chromosome. The integrated provirus behaves like a cellular gene and is transcribed for viral replication into viral mRNAs and a complete (+) sense RNA genome.

Class VII, dsDNA with RNA intermediate (Hepadnaviridae, the plant virus genera *Badnavirus* and *Caulimovirus*). Viruses of this group practice reverse transcription from DNA to RNA. In hepadnaviruses, contrary to retroviruses, this occurs inside the virus particle and not in the cell nucleus. Integration of viral DNA into the host genome is not required for replication. The DNA of caulimoviruses forms a minichromosome in the nucleus of the host cell and is transcribed into an RNA with a terminal redundancy and slightly greater length than the initial genome. This genomic structure serves as the template for the formation of a double-stranded DNA in the cytoplasm.

Following replication, viruses use numerous strategies for gene expression that are influenced by fundamentally different control mechanisms in prokaryotic and eukaryotic hosts. Despite their small genomes, viruses are able to exert quantitative, temporal, and spatial control of gene expression, achieved by positive and negative signals which promote or repress protein syntheses. Viral strategies to overcome small genome sizes include the evolution of overlapping genes and the production of multiple polypeptides from a single messenger RNA (polycistronic mRNA).

Assembly or maturation. This stage corresponds to the formation of mature virus particles and to the acquisition of infectivity. Particle assembly starts when a critical concentration of virus proteins and genomic material is reached. Nucleic acids are packaged into preexisting shells (e.g., in many dsDNA viruses), are coated with capsid proteins (ino- and leviviruses), or may co-assemble with capsid proteins (picornaviruses). Assembly sites differ according to the virus and have some influence on the mechanism of release. For example, picornaviruses, poxviruses, and reoviruses assemble in the cytoplasm, while adenoviruses, papovaviruses, parvoviruses, and herpesvirus capsids assemble in the nucleus. The assembly of retrovirus nucleocapsids occurs on the inner surface of the plasma membrane. Self-assembly seems to be responsible for the formation of many nonenveloped virions. In enveloped viruses, nucleocapsids are formed in the nucleus or the cytoplasm, viral envelope proteins are inserted into cellular membranes (nuclear, Golgi, endoplasmic reticulum, or plasma membranes), and nucleocapsids acquire their envelopes by budding through membranes carrying envelope proteins. Tailed phages have different assembly pathways for heads, tails, and tail fibers. In addition, at least in part of them, their capsid proteins mature by cleavage of precursor proteins. In retroviruses, maturation is completed after the exit of the virus particle from the infected cell.

Release. Novel viruses are released in three ways. (i) Lytic, generally nonenveloped viruses are released by destruction (lysis) of the infected cell. (ii) Enveloped viruses bud directly through the plasma membrane (most enveloped viruses of vertebrates and phages of the *Plasmaviridae* family) or are released by fusion of secretory vesicles loaded with virus particles with the plasma membrane. The budding process provides nonenveloped viral nucleocapsids with an envelope. Herpesvirus capsids are synthetized in the nucleus, acquire a primary envelope from the nuclear membrane and a definitive envelope from the Golgi network, and finally leave the cell by exocytosis. Except for herpesviruses, budding generally does not lead to immediate cell death and allows for some survival of the host. (iii) Filamentous bacteriophages are extruded through bacterial membranes without killing their hosts.

The complement of the productive cycle is the "temperate" cycle, first clearly described in tailed bacteriophages in 1953.[12] Attachment, penetration, and uncoating proceed as in a replicative cycle, but then the viral genome becomes latent and persists in the cell either linearly integrated into the genome or as a circular episome within the cytoplasm. Only DNA and retroviruses, the latter after transcription of their RNA into DNA, are able to establish a temperate state. In tailed phages, the latent genome is called a prophage and the latent state is called lysogeny, meaning that it is able to develop into cell lysis. The latent retrovirus genome is called a provirus. In becoming latent, the viral genome undergoes a symbiotic relationship with the host cell, is replicated more or less synchronously with the host DNA, may prevent superinfections by related viruses, and may confer new properties to its host. The latent state may persist over many cell generations until it breaks down spontaneously or under the action of inducing agents, generally mutagens or carcinogens. The viral genome is then set free and enters a normal replication cycle. The ability to become latent is found in many, very different viruses of vertebrates (*Hepadna-, Herpes-, Papova-, Retroviridae*), bacteria (all three families of tailed phages, the *Fusello-, Lipothrix-, Ino-,* and *Plasmaviridae* families of filamentous or pleomorphic phages), and possibly the *Rhizidiovirus* genus of fungal viruses.[5] This widespread occurrence of latency suggests that the ability to integrate into host DNA arose several times during viral evolution.

Chapter 4

COMPARATIVE DIAGRAMS

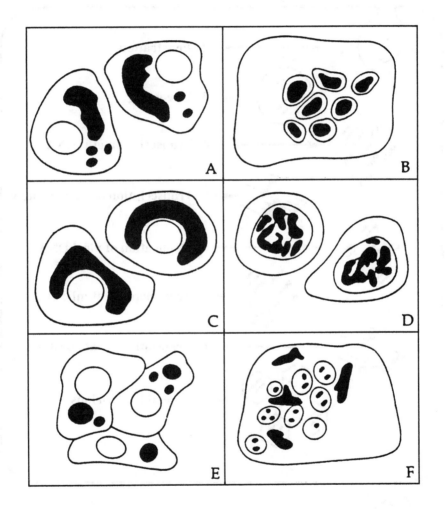

FIGURE 1

Inclusion bodies in virus-infected cells. (A) Cytoplasmic acidophilic Guarnieri bodies caused by vaccinia virus. (B) Nuclear acidophilic inclusions (Cowdry type A) caused by herpes simplex virus; cell fusion produces syncytium. (C) Perinuclear cytoplasmic acidophilic inclusions caused by reoviruses. (D) Nuclear basophilic inclusions caused by adenoviruses. (E) Cytoplasmic acidophilic inclusions caused by rabies virus (Negri bodies). (F) Nuclear and cytoplasmic inclusions caused by measles virus; cell fusion produces syncytia. (From Fenner, F., McAuslan, B.R., Mims, C.A., Sambrook, J., and White, D.O., *The Biology of Animal Viruses,* 2nd ed., Academic Press, New York, 1974, 342. With permission.)

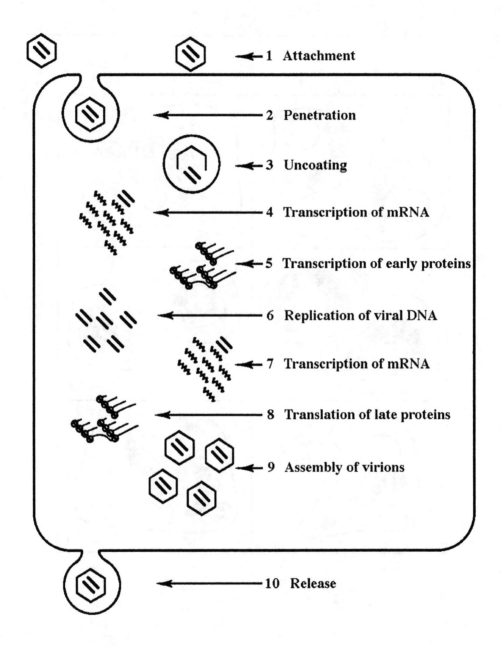

FIGURE 2

General features of the viral replication cycle, using a nonenveloped icosahedral DNA virus as a model. No topographical location, nuclear or cytoplasmic, for any step is implied. One step grades into the next such that, as the cycle progresses, several events are proceeding simultaneously. (Redrawn from Fenner, F.O. and White, D.O., *Medical Virology,* 2nd ed., Academic Press, New York, 1976, 50. With permission.)

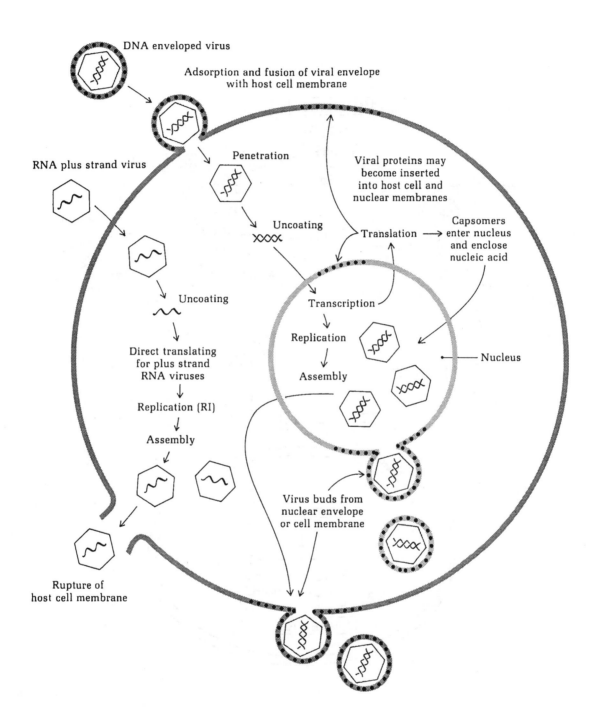

FIGURE 3

Replication of an enveloped DNA virus and an RNA (+) strand virus. Note that with some DNA viruses, such as smallpox virus, viral replication occurs solely in the cytoplasm of the host cell. (From Volk, W.A. and Wheeler, M.F., *Basic Microbiology,* 4th ed., J.B. Lippincott, Philadelphia, 1980, 144. With permission.)

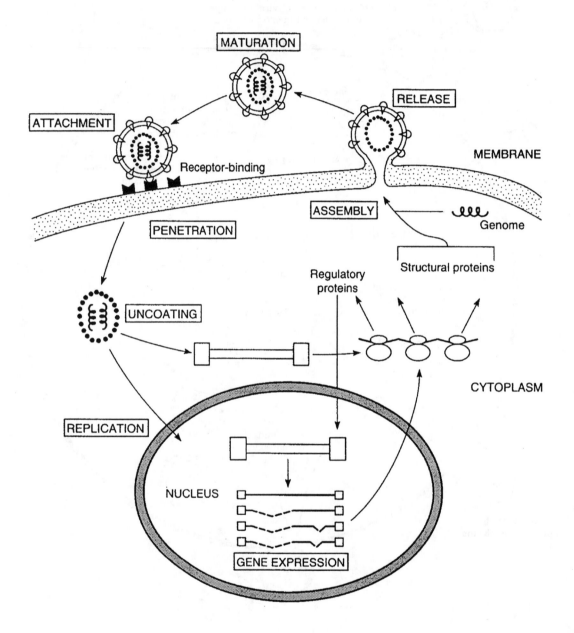

FIGURE 4

Generalized scheme for virus replication. (Reprinted from Cann, A.J., Ed., *Principles of Molecular Virology,* Academic Press, London, 1993, 86. By permission of the publisher Academic Press Ltd., London.)

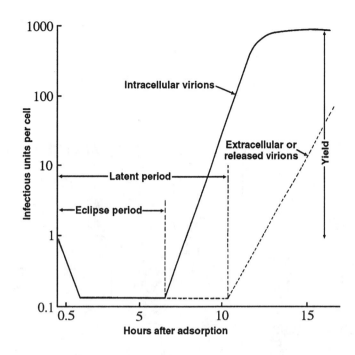

FIGURE 5

Time course of the multiplication cycle of an animal virus that is released from a cell lately and incompletely. In other cases, virions mature as they are released and curves for intracellular and extracellular infectivity correspond. (From Fenner, F., McAuslan, B.R., Mims, C.A., Sambrook, J., and White, D.O., *The Biology of Animal Viruses,* 2nd ed., Academic Press, New York, 1974, 177. With permission.)

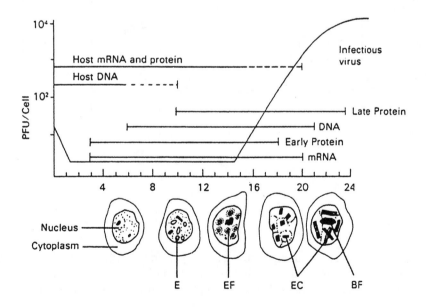

FIGURE 6

Type 5 adenovirus synthesis and concomitant nuclear alterations and changes in host macromolecule synthesis. E, eosinophilic masses; EF, eosinophilic masses with basophilic Feulgen-positive border; EC, eosinophilic crystals; BF, basophilic Feulgen-positive masses; PFU, plaque-forming units. Types 1, 2, and 6 have similar multiplication and nuclear changes. (From Dulbecco, R. and Ginsberg, H.S., in *Microbiology,* 3rd ed., Davis, B.D., Dulbecco, R., Eisen, H.N., and Ginsberg, H.S., Eds., Harper & Row, Hagerstown, MD, 1980, 1047. With permission.)

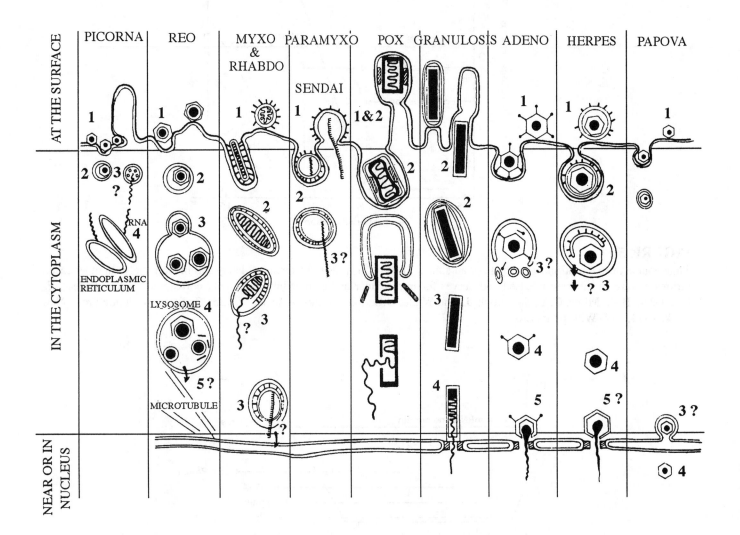

FIGURE 7

Early aspects of animal virus replication, including attachment (1), penetration (2), and uncoating (3, 4, 5). The nucleic acid of the infecting virus must be removed from its capsid before formation of new viruses. In some cases this uncoating begins in phagocytic or pinocytic vesicles and ends in the host cell cytoplasm. (From Dales, S., *Bacteriol. Rev.,* 37, 103, 1973. With permission.)

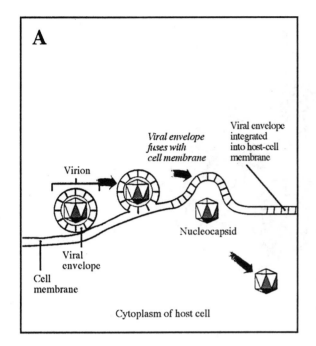

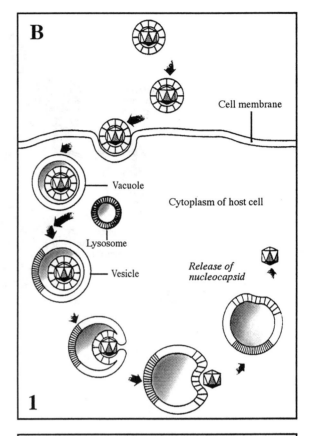

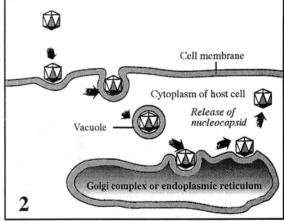

FIGURE 8

(A) Penetration of virion into host cell by fusion of viral envelope and cell membrane. The nucleocapsid is released directly into the cytoplasm. (B) Penetration of virion into host cell by vacuolar ingestion. (1) With an enveloped virus, the viral envelope adsorbs to the cell membrane and the entire virion is engulfed in a vacuole, which then fuses with a lysosome to form a vesicle. The envelope then fuses with the vesicle membrane, releasing the nucleocapsid into the cytoplasm. (2) With a nonenveloped virus, the whole virion is engulfed in a temporary vacuole. The vacuole membrane fuses with an internal membrane system (Golgi complex or endoplasmic reticulum) and releases the particle. (Adapted from Pelczar, M.J., Chan, E.C.S., and Krieg, N.R., *Microbiology — Concepts and Applications,* McGraw-Hill, New York, ©1993, 427. Reproduced with permission of The McGraw-Hill Companies.)

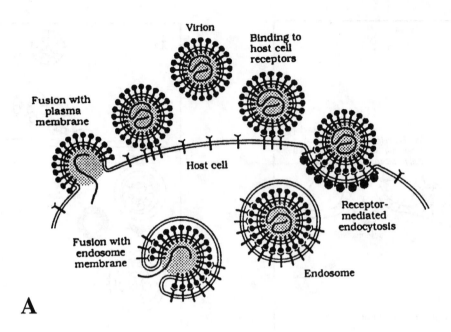

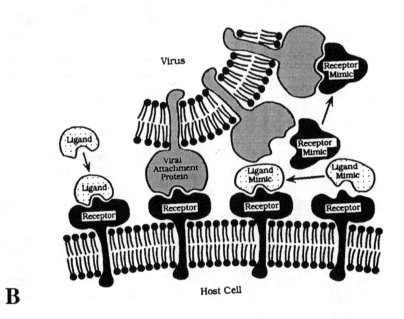

FIGURE 9

(A) Entry of enveloped viruses into cells. The virus particle bears attachment proteins in its membrane, which interact with cell surface molecules (virus receptors) attaching the virion to the cell surface. The viral membrane may then fuse directly with the plasma membrane and the genome is released into the cytoplasm. Alternatively, the virus particle is internalized by adsorptive or receptor-mediated endocytosis and delivered to an endosome. Acidic pH triggers fusion of the viral membrane with the endosome membrane, liberating the genome. (B) Interaction of a virus with a host cell receptor and agents interfering with attachment. The virus binds to the cell via a viral attachment protein (VAP) on its envelope or capsid. The receptor is a normal constituent of the cell surface and may function as a receptor for physiological ligands (e.g., hormones or neurotransmitters). The virus binds to the receptor by a portion of the VAP that mimics a normal ligand structurally or conformationally. Two groups of substances can interfere with the attachment process by blocking the binding of the virus to the receptor: ligand mimics (e.g., antibodies to the binding site of the receptor) and receptor mimics (e.g., synthetic peptides representing the binding domain of the receptor). (Adapted from Lentz, T.L., *J. Gen. Virol.,* 71, 751, 1990. With permission.)

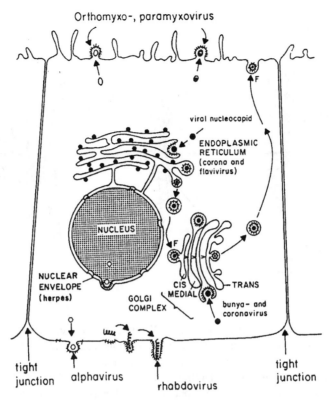

FIGURE 10

Maturation sites of various enveloped viruses. (From Dimmock, N.J. and Primrose, S.B., *Introduction to Modern Virology,* 3rd ed., Blackwell Scientific Publications, Oxford, 1987, 169. With permission.)

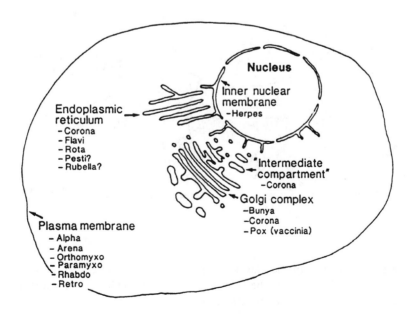

FIGURE 11

Cellular membrane compartments utilized for assembly (budding) of enveloped viruses, with indication of viral families or genera. (From Pettersson, R.F., *Curr. Topics Microbiol. Immunol.,* 170, 67, 1991. © Springer-Verlag. With permission.)

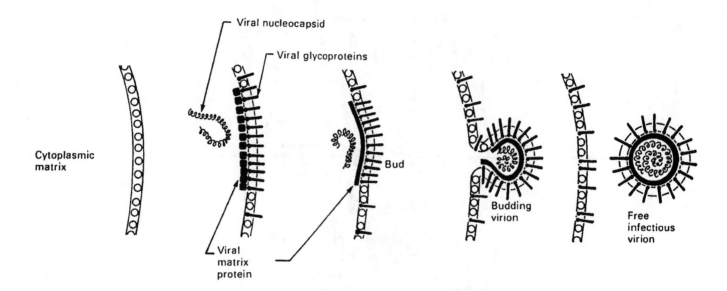

FIGURE 12

Budding of an enveloped virus (orthomyxo- or paramyxovirus). White circles indicate host proteins of the plasma membrane. Black spikes represent viral glycoproteins (peplomers specified by viral genes). Viral matrix (M) protein attaches to the inner surface of the plasma membrane segment that contains viral glycoproteins and appears to serve as recognition site for the nucleocapsid as well as a stabilizing structure. (From Dulbecco, R. and Ginsberg, H.S., in *Microbiology,* 3rd ed., Davis, B.D., Dulbecco, R., Eisen, H.N., and Ginsberg, H.S., Eds., Harper & Row, Hagerstown, MD, 1980, 967. With permission.)

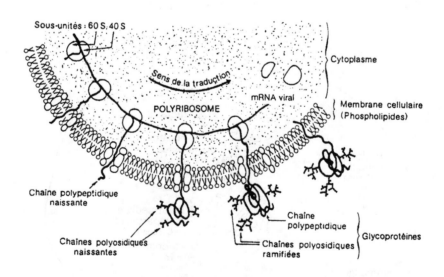

FIGURE 13

Insertion of viral envelope glycoproteins into the plasma membrane of an infected cell. Nascent polypeptide chains are glycosylated and inserted into the cell membrane as they are synthetized. (From Girard, M. and Hirth, L., *Virologie Générale et Moléculaire,* 2nd ed., Doin, Paris, 1989, 369. With permission.)

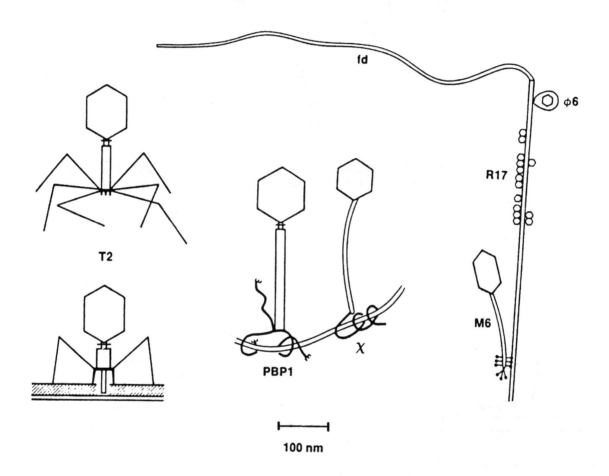

FIGURE 14

Primary adsorption sites of phages. Tailed phages adsorb to the cell wall (coliphage T2), bacterial capsules or stalks (not shown), flagellae (*Bacillus* phage PBP1, enterobacterial phage χ), and pili (*Pseudomonas* phage M6). Various cubic and filamentous phages (inovirus fd, levivirus R17, cystovirus φ6) also adsorb to pili. (From Ackermann, H.-W., in *Virologie Médicale,* Maurin, J., Ed., Flammarion, Paris, 1985, 196. With permission.)

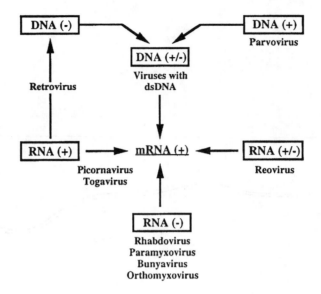

FIGURE 15

Classification of animal viruses by replication mechanism and relationship of genome to mRNA. The category of "+/−" DNA viruses comprises all viruses with double-stranded DNA. (Modified from Girard, M. and Hirth, L., *Virologie Générale et Moléculaire,* 2nd ed., Doin, Paris, 1989, 362. With permission.)

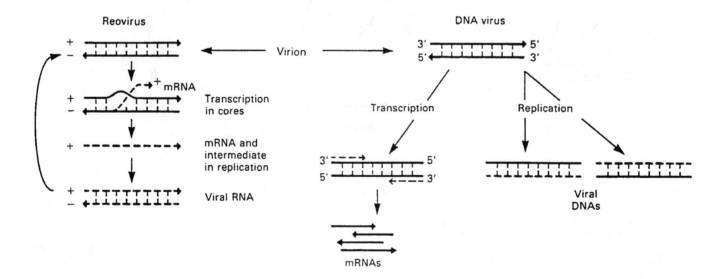

FIGURE 16

Transcription and replication of dsRNA and dsDNA. In the replication of dsRNA, information flows from only one strand (asymmetric) and the molecule is replicated conservatively (i.e., both parental strands are conserved together). In double-stranded viral DNA, information flows from both strands. New strands are indicated by broken lines. (From Dulbecco, R. and Ginsberg, H.S., in *Microbiology,* 3rd ed., Davis, B.D., Dulbecco., R., Eisen, H.N., and Ginsberg, H.S., Eds., Harper & Row, Hagerstown, MD, 1980, 967. With permission.)

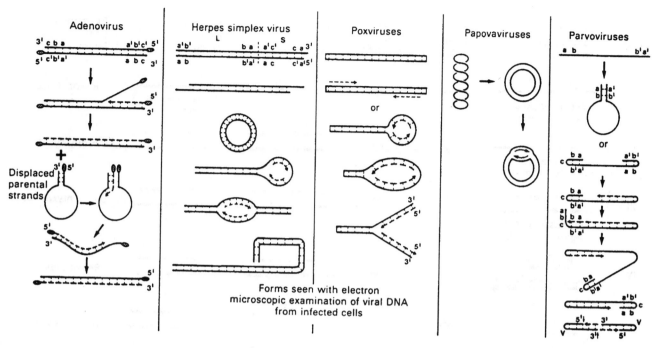

FIGURE 17

Comparative DNA replication in selected viruses. Intermediate stages are proposed from electron microscopic observations and physicochemical determinations made upon DNA replication forms sequentially extracted after infection. (From Dulbecco, R. and Ginsberg, H.S., in *Microbiology,* 3rd ed., Davis, B.D., Dulbecco, R., Eisen, H.N., and Ginsberg, H.S., Eds., Harper & Row, Hagerstown, MD, 1980, 967. With permission.)

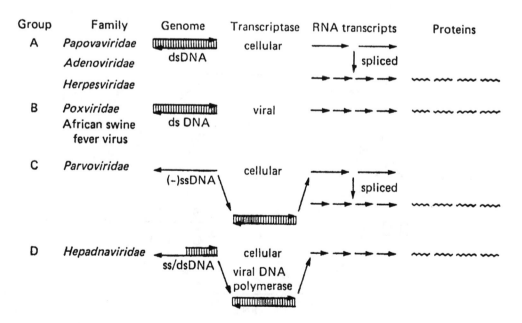

FIGURE 18

Comparative expression strategy of DNA viruses. The sense of each nucleic acid molecule is indicated by arrows (+, to the right; −, to the left). The number of mRNA and protein species is arbitrarily shown as four. (From Fenner, F., Bachmann, P.A., Gibbs, E.P.J., Murphy, F.A., Studdert, M.J., and White, D.O., *Veterinary Virology,* Academic Press, Orlando, 1987, 64. With permission.)

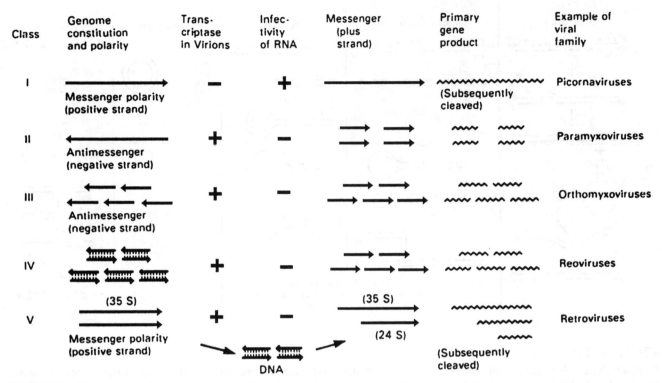

FIGURE 19

Classes of RNA viruses and their primary modes of expression. The numbers of multiple genome pieces, messengers, and gene products are only diagrammatic representations and are not indicative of precise numbers. (From Dulbecco, R. and Ginsberg, H.S., in *Microbiology,* 3rd ed., Davis, B.D., Dulbecco, R., Eisen, H.N., and Ginsberg, H.S., Eds., Harper & Row, Hagerstown, MD, 1980, 967. With permission.)

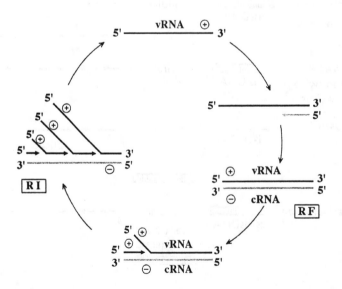

FIGURE 20

Replication of ssRNA viruses. Replication is mediated by specific enzymes (replicases, polymerases). Viral genomic RNA (vRNA) acts as a template for synthesis of complementary RNA strands (cRNA), which in turn are used as templates for synthesis of novel vRNA strands. RF, replicative form; IR, replicative intermediate. (Modified from Girard, M. and Hirth, L., *Virologie Générale et Moléculaire*, 2nd ed., Doin, Paris, 1989, 365. With permission.)

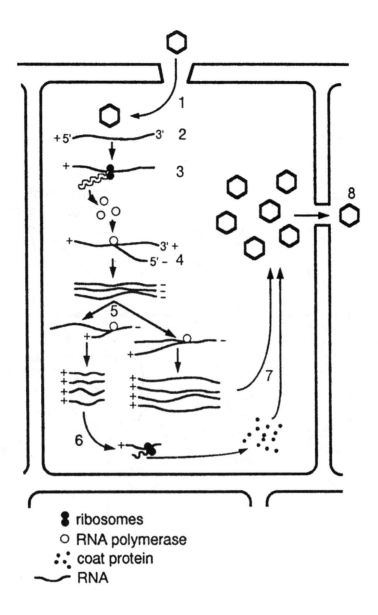

FIGURE 21

Generalized outline for the replication of a small (+) sense ssRNA plant virus:

1. A virus particle enters through a wound made in the cell wall.
2. Viral RNA escapes from the protein coat.
3. Viral RNA becomes associated with host ribosomes and a viral-specific, RNA-dependent RNA polymerase is made.
4. A viral polymerase synthetizes (–) sense genomic lengths of RNA.
5. The polymerase synthetizes genomic-size RNA copies and coat protein mRNAs using an internal initiation site on the genomic (–) sense RNA.
6. Host cell ribosomes synthetize large numbers of coat protein molecules using coat mRNAs.
7. Protein is assembled around genomic RNA molecules to produce progeny viruses, which accumulate in the cell.
8. A few particles migrate to neighboring cells through plasmodesmata.

Note: The features in the diagram are not drawn to scale and there are many variations in detail in the stages illustrated. It is not shown that the processes illustrated often take place in specialized parts of the cytoplasm. (From Matthews, R.E.F., *Fundamentals of Plant Virology,* Academic Press, San Diego, 1992, 106. With permission).

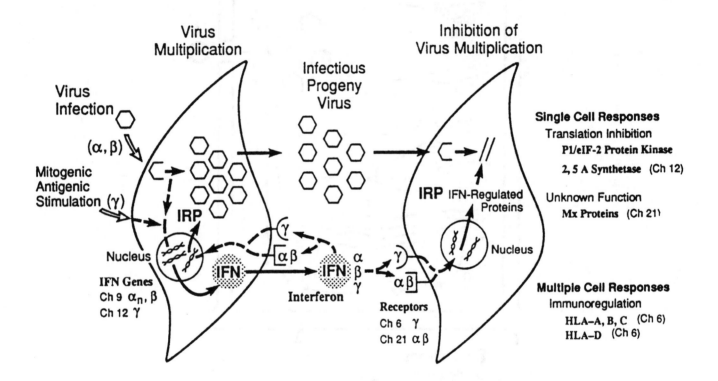

FIGURE 22

The interferon (IFN) system based on the study of human cells. At left, a virus-infected (stimulated) cell produces IFN in response to virus infection (α- and β-IFNs) or mitogens and antigens (γ-IFN). The IFN-treated cell at right responds to the presence of IFN by producing new virus proteins that block virus replication. Some IFN-regulated proteins (IRP), responsible for inhibition of virus replication within single cells (P1eIF-2α protein kinase; 2′,5′-oligoadenylate synthetase, protein Mx) or within whole animals via multiple cell responses (histocompatibility antigens, HLA), are listed on the right. (Modified from Samuel, C.E., in *Encyclopedia of Virology,* Vol. 2, Lederberg, S., Ed.-in-chief, Academic Press, San Diego, 1992, 533. With permission.)

Chapter 5

DNA VIRUSES

I. ssDNA Viruses

5.I.A. INOVIRIDAE

Circular (+) sense ssDNA
Helical, naked
Bacteria

This family includes two genera with about 60 phages. Members of the *Inovirus* genus, exemplified by coliphages fd and f1, are long filaments of 760–2000 × 6–8 nm and infect Gram-negative bacteria (enterobacteria, pseudomonads, xanthomonads, vibrios, *Thermus*). Members of the genus *Plectrovirus* are short rods of 70–280 × 10–14 nm and infect mycoplasmas and spiroplasmas. Coliphages of the *Inovirus* genus adsorb to the tips of pili. The adsorption site of plectroviruses, which infect pilus- and wall-less mycoplasmas, is unknown. Infecting viruses are uncoated at the cell boundary and naked DNA enters the cell. Replication is semi-conservative. The infecting DNA is converted into a double-stranded, supercoiled replicative form (RF). Viral ssDNA is synthetized by a rolling-circle mechanism and is coated again at the cell periphery as particles leave the cell. Particles are extruded without lysis of the host. Some members of the *Inovirus* genus are able to integrate into host DNA and to establish a latent infection.

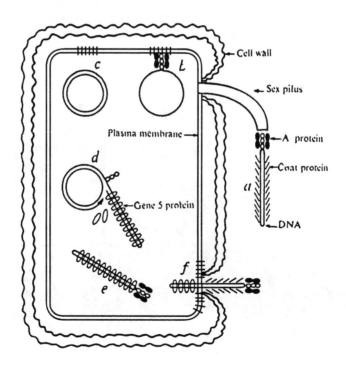

FIGURE 23

Life cycle of fd-type phages of the *Inovirus* genus. (a) The virus attaches to a pilus or other sites of its host by means of the A protein, a minor coat component. (b) Viral DNA and A protein enter the host cell, leaving the major coat protein (gp8) at the plasma membrane. (c) The viral DNA is converted to a duplex form (RFI) which replicates according to the rolling-circle model. (d) The progeny duplex spins off a single-stranded tail that is coated with gp5. (e) The viral DNA is closed to give a circular molecule in a linear DNA–gp5 complex. (f) The DNA exits through the cytoplasmic membrane while gp5 is replaced by novel coat protein. (Reprinted with permission from Marvin, D.A. and Wachtel, E.J., *Nature,* 253, 19, 1975. ©Macmillan Magazines Ltd.)

INOVIRIDAE

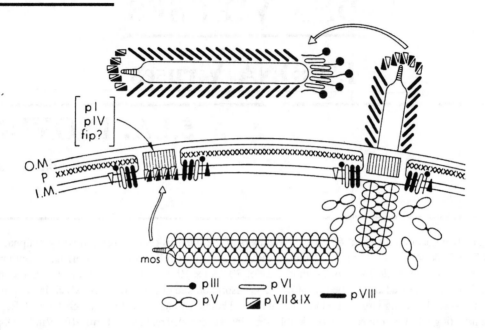

FIGURE 24

Assembly of phage f1. OM, outer membrane; P, periplasm, IM, inner membrane; crosses, peptidoglycan layer; mos, morpho-genetic signal. Gene 5 (V) protein is represented as a dimer. The cell interior with the gp5–DNA complex is shown in the lower part of the figure. (From Webster, J.E. and Lopez, J., in *Virus Structure and Assembly,* Casjens, S., Ed., Jones and Bartlett Publishers, Boston, ©1985, 235. Reprinted with permission.)

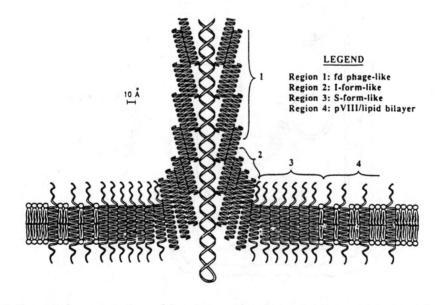

FIGURE 25

Proposed pathway for gp8 (gpVIII) during fd phage penetration and assembly. gp8 passes through several sequential stages and locations including a phage-like structure (region 1), intermediate I and S forms (regions 2 and 3), and a lipid bilayer containing gp8 as a transmembrane protein. gp8 in regions 3 and 4 has a transmembrane helix and nonhelical ends. (Reprinted from Dunker, A.K., Ensign, L.D., Arnold, G.E., and Roberts, L.M., *FEBS Lett.,* 292, 271, ©1991. With kind permission of Elsevier Science–NL, Sara Burgerhartstraat 25, 1055 KV Amsterdam, The Netherlands.)

INOVIRIDAE

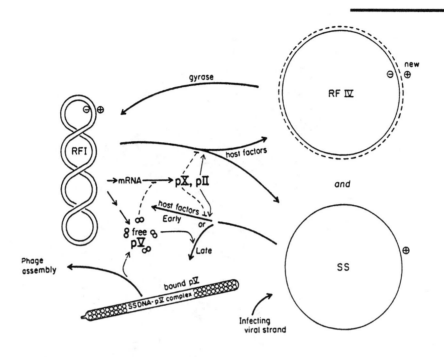

FIGURE 26

Replication of F-specific phage DNA. Thick arrows represent DNA interconversions. Infecting single strands (SS) are converted to supercoiled RFI, which produces a relaxed RF (RF IV) and a viral (+) strand (SS). This strand can be converted to a double strand or sequestered by gene V protein to make a V protein–DNA complex. Proteins X, II, and V are synthetized from RFI. Thin arrows indicate a stimulatory role for the proteins; broken lines indicate indicatory activities. (From Model, P. and Russel, M., in *The Bacteriophages,* Vol. 2, Calendar, R., Ed., Plenum Press, New York, 1988, 375. With permission.)

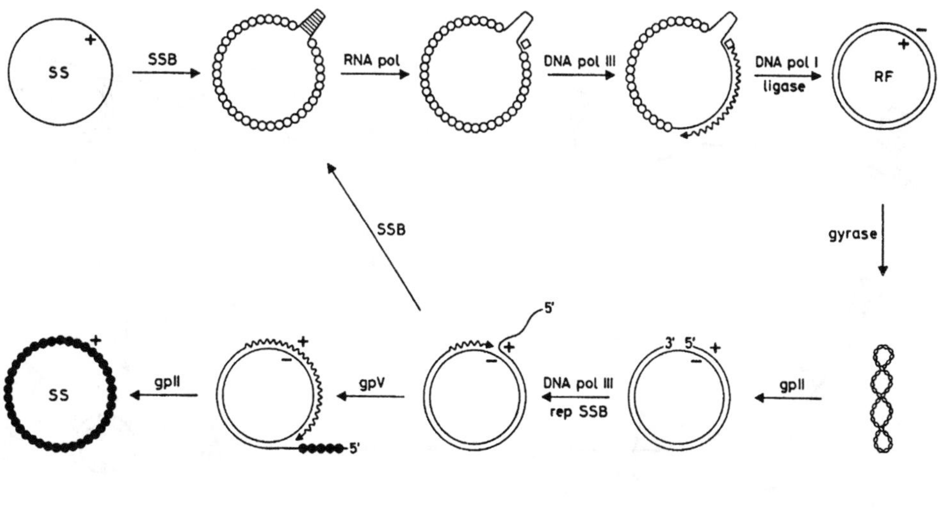

FIGURE 27

The two mechanisms of DNA replication in fd-type phage M13. The upper part shows the discontinuous synthesis of the complementary (–) strand, and the lower part shows the continuous synthesis of the viral (+) strand. Small circles, SSB host protein; small squares, RNA primers; dots, gpV. (From Keppel, F., Fayet, O., and Georgopoulos, C., in *The Bacteriophages,* Vol. 2, Calendar, R., Ed., Plenum Press, New York, 1988, 145. With permission.)

INOVIRIDAE

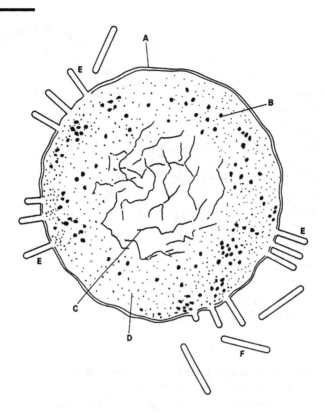

FIGURE 28

A mycoplasma cell (not to scale) with plectrovirus MVL1 in the process of extrusion. (A) Plasmalemma, (B) ribosomes, (C) cellular DNA, (D) cytoplasmic matrix DNA. Viruses are assembled at the cell surface (E) and released into the culture medium (F). (From Horne, R.W., in *The Structure and Function of Viruses,* Edward Arnold, London, 1978, 50. Courtesy of R.W. Horne. With permission.)

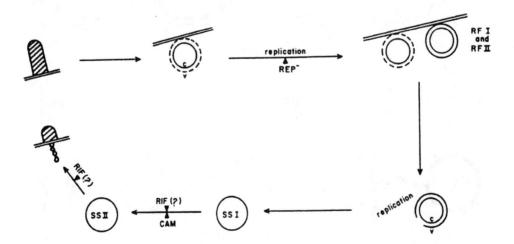

FIGURE 29

Replication of plectroviruses. Dotted lines show parental viral DNA strands and continuous lines show progeny DNA. Some steps are blocked by rifampicin (RIF), chloramphenicol (CAM), or an REP-cell variant. Viral and complementary strands are marked v and c, respectively. Parallel double lines denote the cell membrane. (From Maniloff, J., Das, J., and Christensen, J.R., *Adv. Virus Res.,* 21, 343, 1977. With permission.)

5.I.B. MICROVIRIDAE

Circular (+) sense ssDNA
Cubic, naked
Bacteria

This group of about 40 phages is divided into four genera, exemplified by coliphage φX174 (genus *Microvirus*). Its members infect enterobacteria, bdellovibrios, chlamydias, and spiroplasmas. Particles are icosahedra of 26–32 nm in diameter. Viruses adsorb to cell membranes. Phage DNA enters the cell and is converted into double-stranded, supercoiled RF. Cleavage of this RF by viral A protein starts further RF replication. Viral ssDNA is synthetized via a rolling-circle mechanism as in inoviruses. In the third stage, DNA synthesis is coupled to the morphogenetic pathway of the phage. Novel viral DNA is packaged into procapsids, which mature through loss of two nonstructural proteins. Progeny viruses are released by cell lysis.

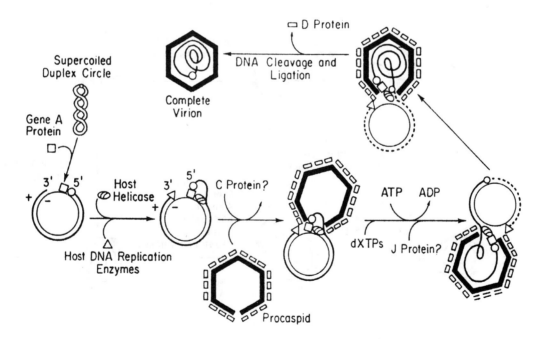

FIGURE 30

Replication and packaging of φX174 DNA. The viral (+) strand is a bold line and the complementary (–) strand is a fine line. Newly synthetized strands are indicated by dashed lines. The solid circle on the viral strand represents the A protein cleavage site. (From Casjens, S., in *Virus Structure and Assembly,* Casjens, S., Ed., Jones and Bartlett Publishers, Boston, ©1985, 75. Reprinted with permission.)

MICROVIRIDAE

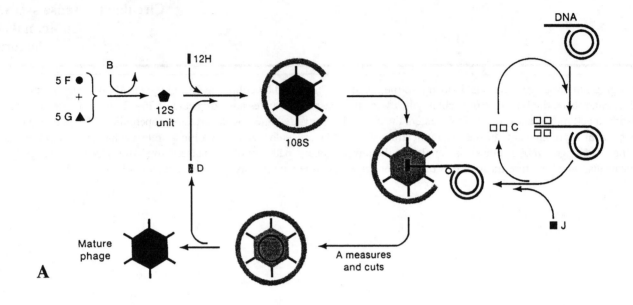

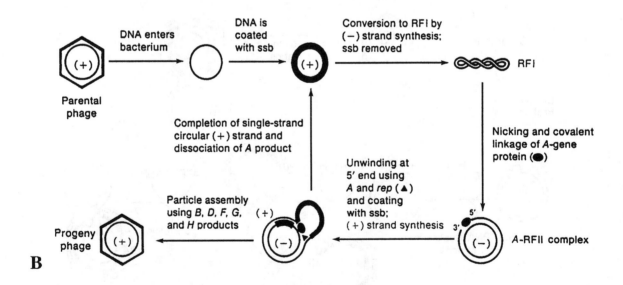

FIGURE 31

(A) Assembly of phage φX174. Protein B catalyzes aggregation of five molecules of the major capsid protein (F), with five molecules of the spike protein (G) to form a 12S unit. Protein D may provide a scaffolding function to form the 108S particle. Protein C facilitates encapsidation of DNA. Protein A measures a unit length of DNA and forms the circle. The role of protein J is uncertain. (B) φX174 replication. Parental DNA is converted to double-stranded supercoiled RFI. Gene A protein then makes a nick and the looped rolling-circle mechanism begins, causing displacement of the progeny (+) strand made during formation of RFI. In the presence of phage-encoded maturation and coat proteins, progeny (+) strand circles are encapsidated rather than converted to RFI. Except for gene A proteins, all proteins (ssb, rep) used in replication are bacterial. (Adapted from Freifelder, D., *Molecular Biology, A Comprehensive Introduction to Prokaryotes and Eukaryotes,* Science Books International, Boston, 1983, 653 and 654. ©1983 Boston: Jones and Bartlett Publishers. Reprinted with permission.)

MICROVIRIDAE

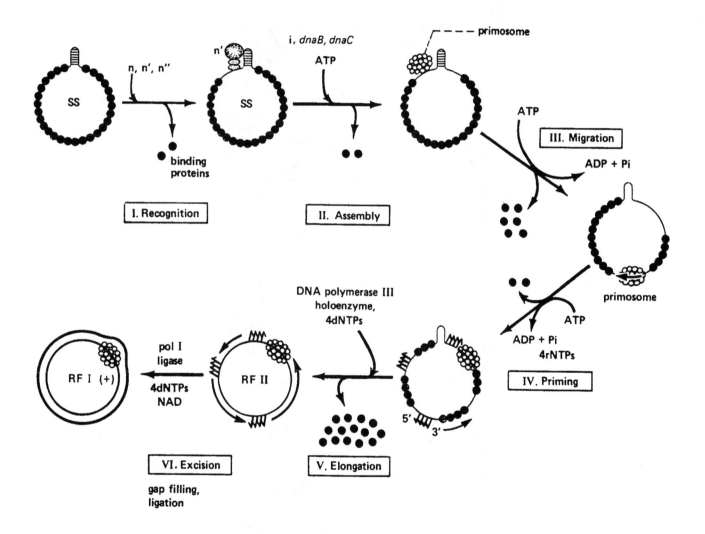

FIGURE 32

Stage I DNA replication. (I) Recognition of primosome site by protein n′. (II) Assembly of primosome. (III) Migration of primosome. (IV) Priming. (V) DNA elongation by DNA polymerase III. (VI) Excision of primers, gap filling, and closure. n′, *dnaB,* and *dnaC* are host proteins. (From Arai, K.-L., Low, R., Kobori, J., Shlomai, J., and Kornberg, A., *J. Biol. Chem.,* 256, 5273, 1981. With permission.)

MICROVIRIDAE

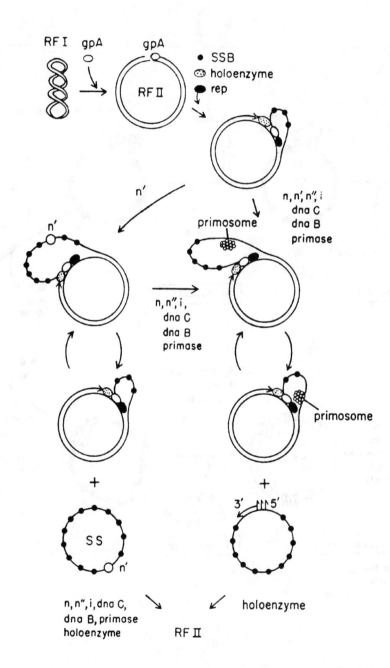

FIGURE 33

Stage II DNA replication. Viral strand replication requires several host proteins (SSB, polymerase III holoenzyme, rep) and proceeds through a rolling-circle mechanism. (From Hayashi, M., Aoyama, A., Richardson, D.E., and Hayashi, M.N., in *The Bacteriophages,* Vol. 2, Calendar, R., Ed., Plenum Press, New York, 1988, 1; modified from Ref. 40. With permission.)

5.I.C. PARVOVIRIDAE

Linear (+) or (–) sense dsDNA
Cubic, naked
Vertebrates, invertebrates

Viruses are widely distributed in animals. Members of the *Parvovirinae* subfamily (three genera) infect vertebrates. Among these, the genus *Dependovirus* includes defective viruses that require coinfection by adenoviruses or herpesviruses for replication and are able to integrate into the host chromosome. The *Densovirinae* subfamily (three genera) includes viruses of insects and possibly crustaceans. Genomes are linear ssDNA with (–) (complementary to viral mRNA) or (+) polarity. Some viruses encapsidate (+) or (–) strands only; others encapsidate (+) and (–) strands in equivalent or different proportions. The processes of adsorption, uncoating, and replication are imperfectly understood. The replication model in Figure 35 is for a dependovirus and is an example only. Viral replication occurs in the nucleus and proceeds via double-stranded or concatemeric intermediates.

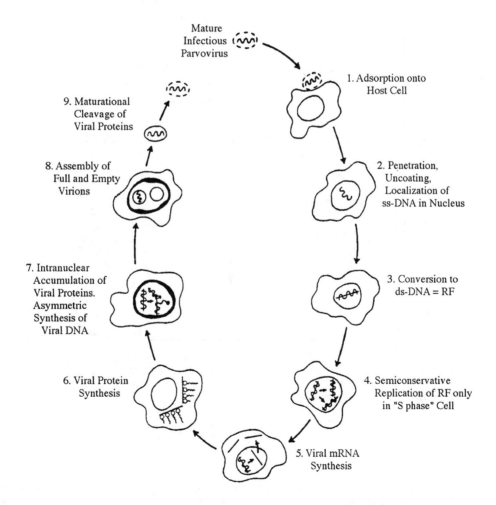

FIGURE 34

Replication of nondefective parvoviruses. (From Toolan, H.W. and Ellem, K.O.A., in *Virology and Rickettsiology,* Vol. I, Part 2, Hsiung, G.-D. and Green, R.H., Eds., CRC Press, Boca Raton, FL, 1978, 3. With permission.)

PARVOVIRIDAE

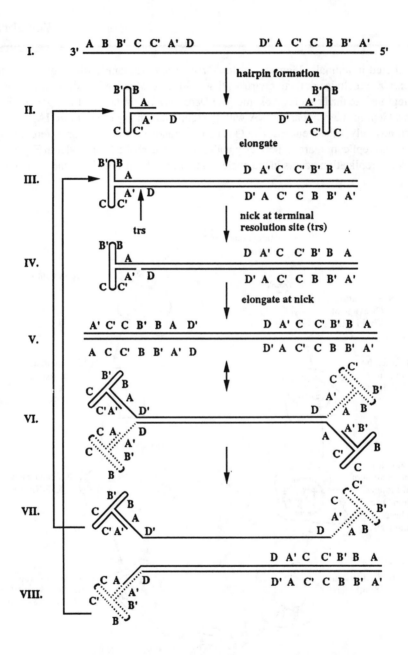

FIGURE 35

Replication of AAV (adeno-associated virus). The terminal repeats are self-complementary and capable of forming hairpins (II). This allows for self-primed DNA synthesis from the 3'OH-group. The site- and strand-specific nick shown in IV is made by Rep proteins 68 or 78, which also possess helicase activity for conversion (isomerization) of structure V to VI. Structure VII is encapsidated during strand displacement (resulting in net virion production) or enters the template amplification pathway (VIII). (Redrawn from Berns, K.I. et al., in *Virus Taxonomy, Sixth Report of the International Committee on Taxonomy of Viruses,* Murphy, F.A. et al., Eds., Springer, Vienna, *Arch. Virol.,* 10, 169, 1995. © Springer-Verlag. With permission.)

II. dsDNA Viruses with Cubic or Helical Symmetry

5.II.A. ADENOVIRIDAE

Linear dsDNA
Cubic, naked
Vertebrates

The name is derived from adenoid tissue, from which the first representatives were isolated. They comprise mammalian and avian viruses with numerous serotypes (genera *Mastadenovirus* and *Aviadenovirus*). Particles are nonenveloped icosahedra 80–110 nm in diameter. Viruses enter cells directly or by receptor-mediated endocytosis. DNA replication occurs in the nucleus, is semiconservative and protein-primed, and proceeds by strand displacement (in a rolling-circle mechanism?). Novel DNA enters preformed capsids. Progeny viruses are assembled in the nucleus and released by cell lysis.

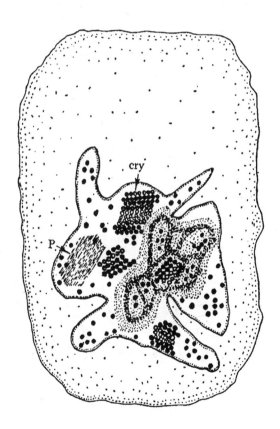

FIGURE 36
Cell showing adenovirus replication. Virions are found scattered or in crystalline arrays (cry) in a deformed nucleus. P, protein crystal of unknown nature. The diagram shows that many features of adenovirus replication were known at an early date. (From Bernhard, W., in *Ciba Foundation Symposium on Cellular Injury,* De Reuck, A.V.S. and Knight, J., Eds., J. & A. Churchill, London, 1964, 209. With permission.)

ADENOVIRIDAE

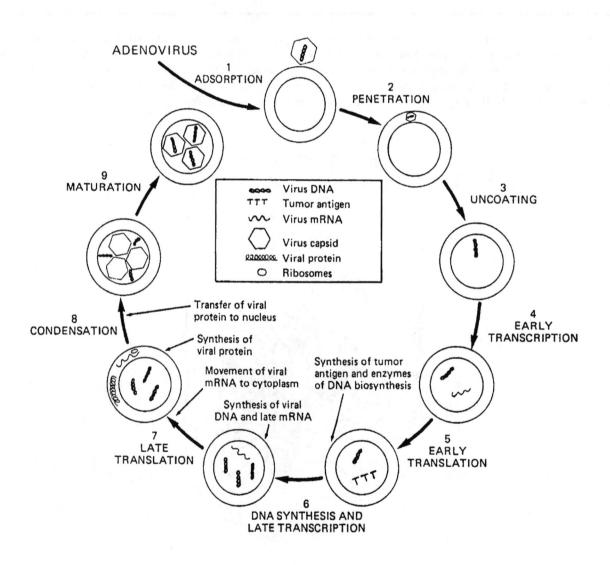

FIGURE 37

Steps in the replication of adenoviruses. (From Jawetz, E., Melnick, J.L., and Adelberg, E.A., *Review of Medical Microbiology*, 15th ed., Lange Medical Publications, Palo Alto, CA, 1982, 336. © Appleton & Lange. With permission.)

ADENOVIRIDAE

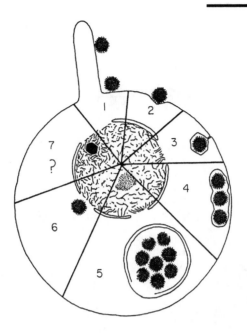

FIGURE 38

Adenovirus 7 replication in HeLa cells. Stages 1 and 2 were observed 1 h after adding virus to cultures; stages 3 to 6 were found 2 h after infection. (The existence of stage 7 was not confirmed in subsequent studies; authors' note, see Ref. 46.) (Reproduced from Dales, S., *J. Cell Biol.,* 13, 303, 1962. By copyright permission of the Rockefeller University Press.)

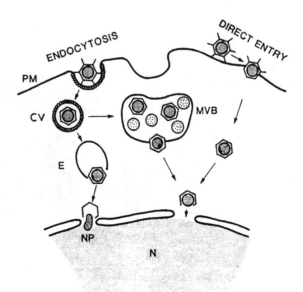

FIGURE 39

Two modes of adenovirus entry into cells: receptor-mediated endocytosis and direct entry. PM, plasma membrane; CV, coated vesicle with adenovirus; E, endosome; MVB, multivesicular body; N, nucleus; NP, nuclear pore. Viruses penetrate the endosomal membrane by an unknown mechanism, migrate toward the nucleus, and are uncoated close to nuclear pores. In some species, particles appear in lysosomes, but this is not typical. (From Nermut, M.V., in *Animal Virus Structure,* Nermut, M.V. and Steven, A.C., Eds., Elsevier, Amsterdam, 1987, 373. With permission.)

ADENOVIRIDAE

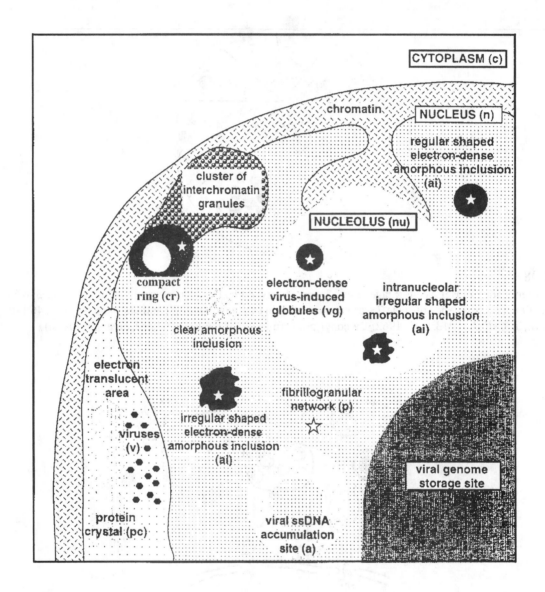

FIGURE 40

Alterations of host cell nucleus after adenovirus infection. Electron-dense amorphous inclusions, either round or less regular, were found in both the nucleoplasm and the nucleolus. They were tentatively given a single name (ai), their origin being presently unclear. Stars indicate structures labeled with antibodies against adenovirus protein IVa2. (From Lutz, P., Puvion-Dutilleul, F., Lutz, Y., and Kedinger, C., *J. Virol.*, 70, 3449, 1996. With permission.)

ADENOVIRIDAE

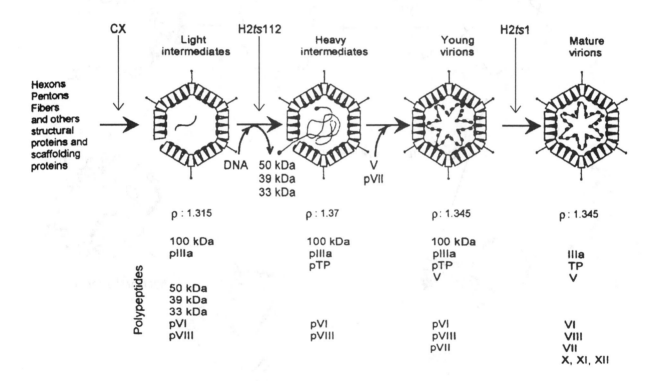

FIGURE 41

Adenovirus assembly pathway. Structural proteins are synthetized and transported to the nucleus in the form of capsomers. Empty capsids are formed from hexons or nanomers of hexons around scaffolding protein(s). The main assembly intermediates contain some proteins absent in the mature virion. Further steps include the release of scaffolding protein(s), encapsidation of DNA, and proteolytic cleavage of at least five precursor polypeptides in the young virion. Polypeptides only present in some structures are indicated below. CX, cycloheximide; H2ts112 and H2ts1, defective mutants. (From D'Halluin, J.C., *Curr. Topics Microbiol. Immunol.*, 109(1), 47, 1995. ©Springer-Verlag, Berlin. With permission.)

ADENOVIRIDAE

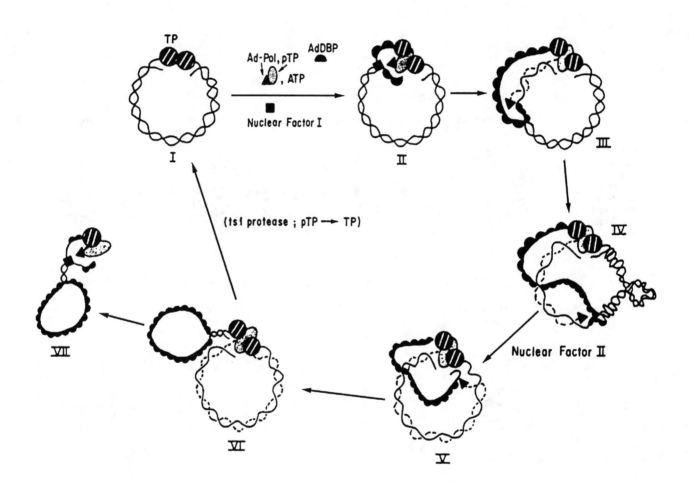

FIGURE 42

Adenovirus DNA replication. The proposed model is a modified rolling-circle mechanism in which single-stranded displacement occurs concomitantly with DNA synthesis. After initiation and elongation of structure I, a displacement reaction produces structure II. A replication intermediate (III) is formed with continued elongation. The absence of topoisomerase I causes the accumulation of structure IV; its presence permits the chains to become further elongated. Displacement of an intact single strand from structure VI can lead to the panhandle structure VII (not yet seen). (From Friefeld, B.R., Lichy, J.H., Field, J., Gronostajski, R.M., Guggenheimer, R.A., Krevolin, M.D., Nagata, K., Hurwitz, J., and Horwitz, M.S., *Curr. Topics Microbiol. Immunol.,* 110(2), 221, 1984. © Springer-Verlag, Berlin. With permission.)

ADENOVIRIDAE

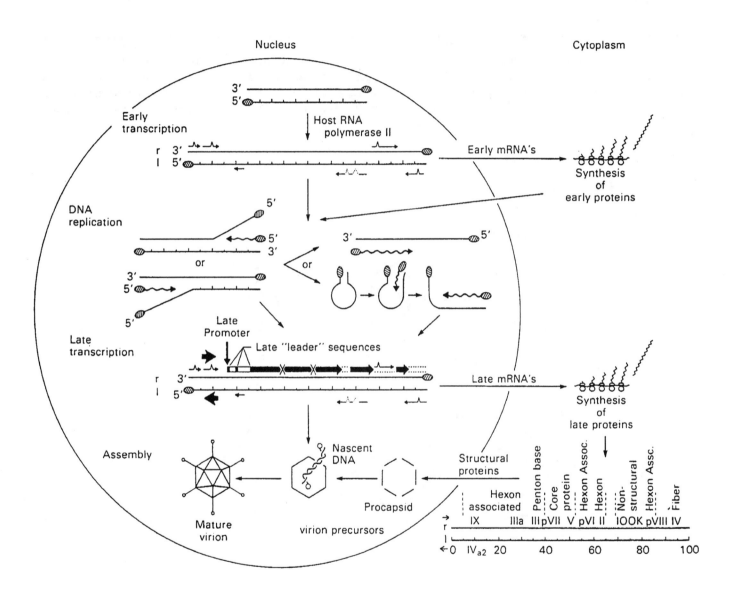

FIGURE 43

Biosynthetic events in the multiplication of adenovirus type 2 (used as a model). Early mRNAs are transcribed from five separate regions of the genome. DNA replication is shown as semi-conservative and asymmetric. A mechanism for replication of displaced single strands, using the inverted terminal repetition to form a circle-like intermediate, is suggested and the map position of late viral proteins is shown. (From Dulbecco, R. and Ginsberg, H.S., in *Microbiology,* 3rd ed., Davis, B.D., Dulbecco, R., Eisen, H.N., and Ginsberg, H.S., Eds., Harper & Row, Hagerstown, MD, 1980, 1047. With permission.)

5.II.B. AFRICAN SWINE FEVER GROUP

Linear dsDNA
Cubic, naked
Vertebrates

This group, which has the status of a "floating" genus, consists of a single enveloped virus 170–125 nm in diameter with an icosahedral capsid. The virus was first classified among iridoviruses and resembles poxviruses in genome structure and strategy of replication. It enters cells by receptor-mediated endocytosis. The DNA replicates in the nucleus under production of concatemers and apparently enters preformed capsids. Progeny viruses are released by cell lysis or budding through the plasma membrane.

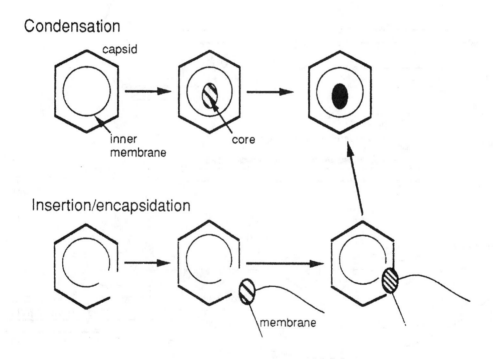

FIGURE 44

Two possible mechanisms for particle maturation. Electron-microscopic observations suggest that empty particles assemble within intracytoplasmic "virus factories" and that nucleoprotein material enters empty capsids. The latter are then sealed, mature into "full" particles, and leave the virus factory before being released by budding. There is no evidence for formation and condensation of nucleoprotein within empty capsids. (From Brookes, S.M., Dixon, L.K., and Parkhouse, R.M.E., *Virology,* 224, 84, 1996. With permission.)

5.II.C. BACULOVIRIDAE

Circular supercoiled dsDNA
Enveloped, helical
Invertebrates

Viruses infect insects and crustaceans (shrimps) and are generally occluded in virus-coded protein crystals, called polyhedra or granules. Particles consist of an envelope and cylindrical nucleocapsids 220–400 × 50 nm. The family includes the genera *Polyhedrovirus* (large polyhedra with 20–200 viruses with one or more nucleocapsids) and *Granulovirus* (small granules containing one virus with a single nucleocapsid). Insects are infected in the larval stage. Inclusion bodies are ingested by prospective hosts and dissolved in the gut. Free viruses enter cells by fusion or endocytosis and replicate and assemble in the nucleus. Progeny DNA enters preformed capsids. Novel viruses acquire an envelope by budding or *de novo* synthesis, spread further within the host, or are again incorporated into inclusion bodies.

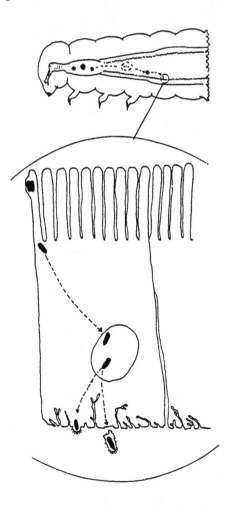

FIGURE 45

The infectious cycle of nuclear polyhedrosis and granulosis viruses starts when viral inclusion bodies are ingested by susceptible larvae. Viruses are released into the midgut after dissolution of inclusion bodies by alkaline gut juices. The released viruses infect the columnar cells of the midgut and replicate in their nuclei. Progeny viruses bud through the basement membrane into the hemocoel, causing a systemic infection. (From Volkman, L.E., *Curr. Topics Microbiol. Immunol.*, 131, 103, 1986. ©Springer-Verlag, Berlin. With permission.)

BACULOVIRIDAE

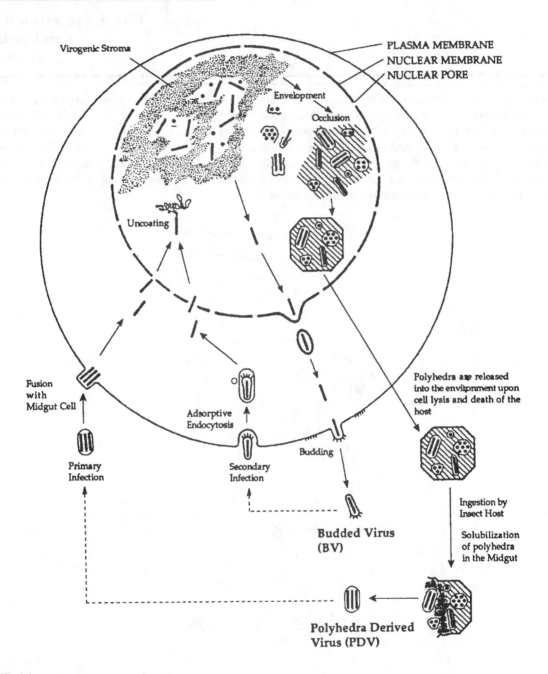

FIGURE 46

Infectious cycle of a nuclear polyhedrosis virus. Polyhedra are ingested by a susceptible insect and solubilized within the insect midgut. Polyhedra-derived viruses (PDV) enter midgut cells by fusion with microvilli. Uncoating of DNA and viral replication take place in the nucleus. Progeny nucleocapsids assemble within and around a virogenic stroma. Some progeny nucleocapsids leave the nucleus and bud into the hemocoel, acquiring an envelope in the process. These budded virions (BV) are able to initiate a systemic infection. Other progeny nucleocapsids become enveloped within the nucleus and are occluded within polyhedra. Upon insect death and cell lysis, the polyhedra are released into the environment. (From Blissard, G.W. and Rohrmann, G.F., *Annu. Rev. Entomol.*, 35, 127, 1990. With permission from the *Annual Review of Entomology,* © 1990, by Annual Reviews Inc.)

BACULOVIRIDAE

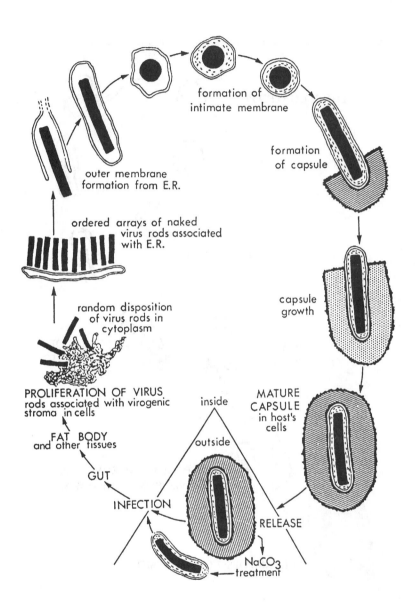

FIGURE 47

Assembly of a granulosis virus. Granules containing a single virus only are digested in the gut and release viruses which infect the fat body and other tissues. Progeny nucleocapsids are assembled within the virogenic stroma of the nucleus, migrate into the cytoplasm, form ordered arrays in the endoplasmic reticulum (ER), acquire an envelope there, and are finally provided with a capsule. (From Arnott, H.J. and Smith, K.M., *J. Ultrastruct. Res.*, 21, 251, 1967. With permission.)

BACULOVIRIDAE

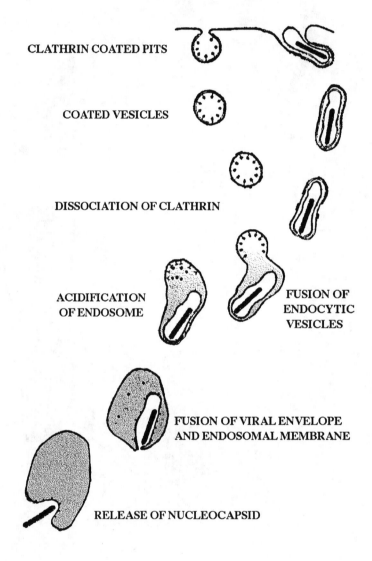

CLATHRIN COATED PITS

COATED VESICLES

DISSOCIATION OF CLATHRIN

ACIDIFICATION
OF ENDOSOME

FUSION OF
ENDOCYTIC
VESICLES

FUSION OF VIRAL ENVELOPE
AND ENDOSOMAL MEMBRANE

RELEASE OF NUCLEOCAPSID

FIGURE 48

Entry of baculoviruses through endocytosis. Viruses (and macromolecules) bind to the cell surface and gather into invaginations of the plasma membrane called "pits." These pits, coated with the protein clathrin, pinch off to become coated vesicles. The clathrin dissociates from the vesicles and two or more of the latter may fuse to become an endosome. The pH of the endosome decreases, the viral envelope and the endosomal membrane fuse, and the nucleocapsid is released into the cytoplasm. (From Volkman, L.E., *Curr. Topics Microbiol. Immunol.,* 131, 103, 1986. ©Springer-Verlag, Berlin. With permission.)

BACULOVIRIDAE

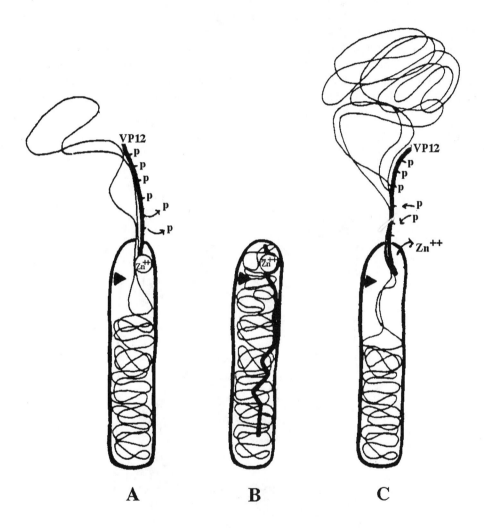

FIGURE 49

Model of *Plodia interpunctella* granulosis virus nucleocapsid during assembly, the fully assembled state, and uncoating. (A) DNA complexed with VP12 enters preassembled capsids through a cap structure. VP12, initially phosphorylated when it binds to viral DNA, becomes dephosphorylated by cellular phosphatases as it enters the capsid, causing condensation and compaction of the DNA. A kinase (arrowhead) binds to the capsid and Zn^{++} is bound by VP12. (B) Stable nucleocapsid containing Zn^{++} and kinase. (C) Uncoating of DNA starts with chelation of Zn^{++}. The kinase phosphorylates VP12, causing relaxation of the DNA protein core and ejection of DNA. (From Funk, C.J. and Consigli, R.A., *Virology,* 193, 396, 1993. With permission.)

5.II.D. HERPESVIRIDAE

Linear dsDNA
Cubic, enveloped
Vertebrates

Members of this large family occur in all branches of vertebrates. They comprise many human pathogens and belong to three subfamilies according to DNA content and structure (*Alpha-*, *Beta-*, and *Gammaherpesvirinae*). Members tend to cause latent infections. Particles are enveloped, 120–200 nm wide, and contain an icosahedral capsid 100–110 nm in diameter. Viral capsids enter cells after fusion of the envelope with the plasma membrane. Viral replication and most of the assembly take place in the nucleus. Infecting DNA circularizes and replicates by a rolling-circle mechanism, yielding concatemers. Progeny DNA enters preformed capsids which leave the nucleus by budding, acquiring an envelope in the process, and transit through the Golgi network. Mature viruses leave the cell by exocytosis.

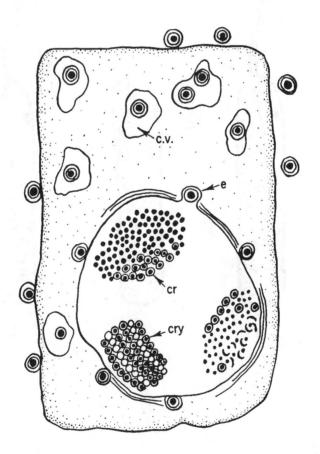

FIGURE 50

Cell showing herpesvirus replication. Viral DNA and capsids form in the nucleus. Capsids are sometimes arranged in a crystalline array (cry). They acquire an envelope (e) by budding through the nuclear membrane and are then found in cytoplasmic vacuoles (c.v.) from which they are eventually released by egestion. (From Bernhard, W., in *Ciba Foundation Symposium on Cellular Injury,* De Reuck, A.V.S. and Knight, J., Eds., J. & A. Churchill, London, 1964, 209. With permission.)

HERPESVIRIDAE

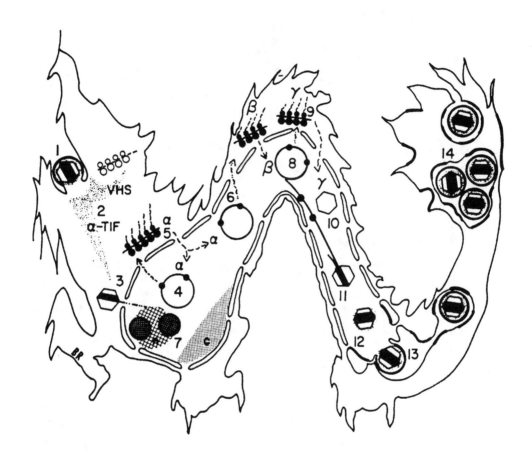

FIGURE 51

Replication of herpes simplex virus, prototype of α herpesviruses. (1) The virus attaches to the cell surface and its envelope fuses with the plasma membrane. (2) This fusion releases two proteins from the virion, VHS (which shuts off protein synthesis) and α-TIF (α gene *trans*-inducing factor). (3) The capsid is transported to a nuclear pore; viral DNA is released into the nucleus and circularizes. (4) Transcription of α genes by cellular genes is induced by α-TIF. (5) Five α mRNAs are transported into the cytoplasm and translated. (6) A new round of transcription results in the synthesis of β proteins. (7) Chromatin (c) is degraded and displaced toward the nuclear membrane, whereas the nucleoli (hatched) become disaggregated. (8) Viral DNA is replicated by a rolling-circle mechanism that yields head-to-tail concatemers of viral DNA. (9) A third round of transcription yields γ proteins, primarily structural proteins of the virus. (10) Capsid proteins form empty capsids. (11) Unit-length viral DNA is cut from concatemers and packaged into preformed capsids. (12) The capsids acquire a new protein. (13) Viral glycoproteins and tegument proteins accumulate and form patches in cellular membranes. Capsids containing DNA and the additional protein attach to these plaques and are enveloped. (14) Enveloped capsids accumulate in the endoplasmic reticulum and leave the cell. (From Roizman, B. and Sears, A.E., in *Virology,* 2nd ed., Fields, B.N. and Knipe, D.M., Eds.-in-chief, Raven Press, New York, 1990, 1795. With permission.)

HERPESVIRIDAE

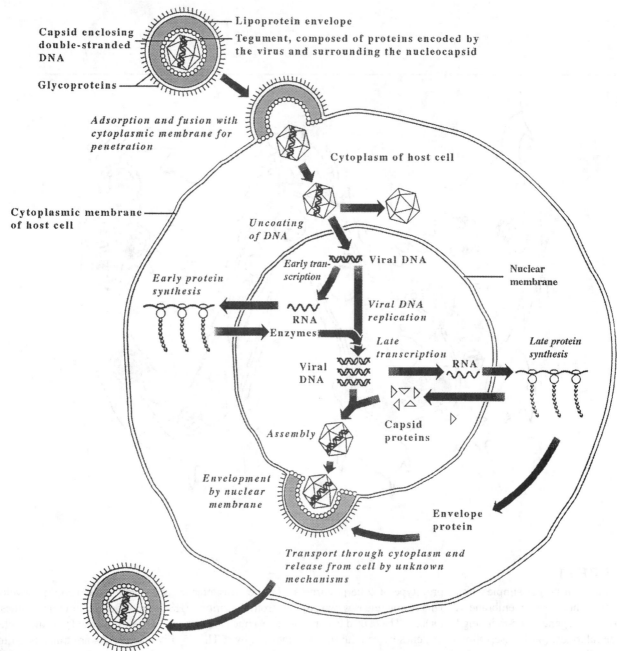

FIGURE 52

Replication of herpes simplex virus. Specific glycoproteins in the viral envelope are essential for adsorption on host cell receptors in the cytoplasmic membrane. Envelope and cell membrane fuse and the viral nucleocapsid is released into the cytoplasm. Viral DNA is then uncoated and transported to the nucleus. Early transcription and mRNA are apparently catalyzed by host enzymes. The resulting early viral enzymes are used in DNA replication. Further RNA transcripts are responsible for synthesis of viral capsid and envelope proteins as well as glycoproteins in the nuclear membrane. The structural proteins enter the nucleus to participate in the assembly of novel viral nucleocapsids. The latter are enveloped by budding through the nuclear membrane and complete viruses are released by unknown mechanisms. (Modified from Pelczar, M.J., Chan, E.C.S., and Krieg, N.R., *Microbiology — Concepts and Applications,* McGraw-Hill, New York, © 1993, 428. Reproduced with permission of The McGraw-Hill Companies.)

HERPESVIRIDAE

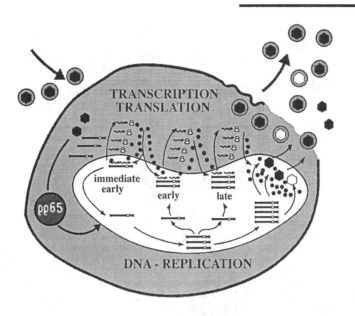

FIGURE 53

Replication of human cytomegalovirus (β herpesviruses). After penetration into the host cell, the virus disintegrates. Viral DNA and tegument protein p65 are transported to the cell nucleus. Viral genes are transcribed in a cascade fashion. Expression of immediate–early genes is a prerequisite for expression of early and late genes. After replication of viral DNA and synthesis of viral proteins, viral capsids are assembled in the nucleus and released from the cell. (From Jahn, G. and Plachter, B., *Intervirology,* 35, 60, 1993. Reproduced with permission of S. Karger AG, Basel.)

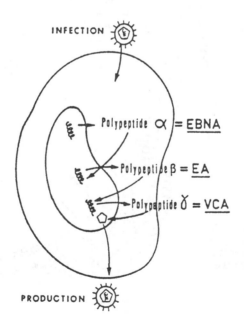

FIGURE 54

Kinetics of synthesis of viral polypeptides in herpesviruses, applied here to the Epstein–Barr virus (γ herpesviruses). EBNA represents the nuclear antigen, EA the early antigen, and VCA the viral capsid or viral structural antigen. (From De Thé, G. and Lenoir, G., in *Comparative Diagnosis of Viral Diseases,* Vol. I, Part A, Kurstak, E. and Kurstak, C., Eds., Academic Press, New York, 1977, 195. With permission.)

HERPESVIRIDAE

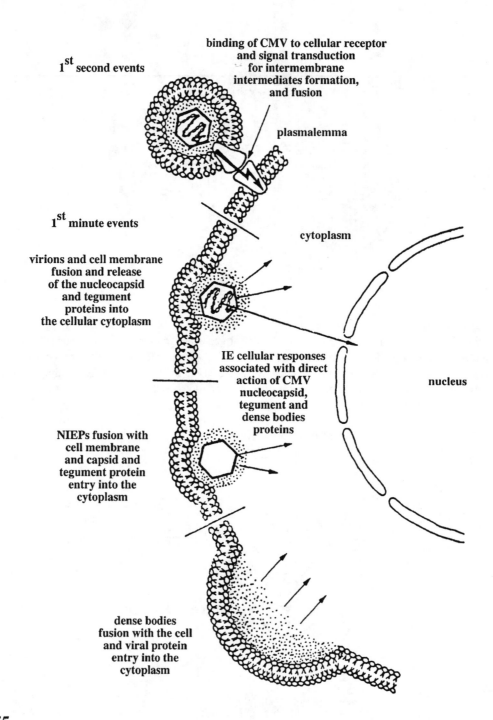

1ˢᵗ second events

binding of CMV to cellular receptor and signal transduction for intermembrane intermediates formation, and fusion

plasmalemma

1ˢᵗ minute events

virions and cell membrane fusion and release of the nucleocapsid and tegument proteins into the cellular cytoplasm

cytoplasm

IE cellular responses associated with direct action of CMV nucleocapsid, tegument and dense bodies proteins

nucleus

NIEPs fusion with cell membrane and capsid and tegument protein entry into the cytoplasm

dense bodies fusion with the cell and viral protein entry into the cytoplasm

FIGURE 55

Interactions of human cytomegalovirus with the host cell during the first 60 seconds of virus adsorption. Particle attachment is mediated by fine bridges. Numerous virions have fused with cell membranes and nucleocapsids have entered the cytoplasm. Noninfectious enveloped particles (NIEPs) with translucent cores and dense bodies (a by-product of virus synthesis) enter the cell the same way as complete virions. (From Topilko, A. and Michelson, S., *Res. Virol.,* 145, 75, 1994. With permission.)

HERPESVIRIDAE

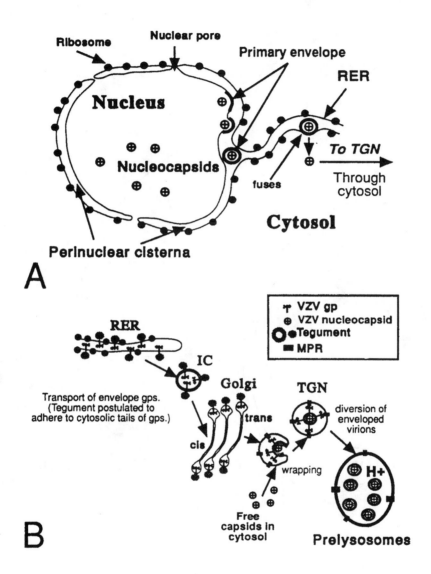

FIGURE 56

Assembly of varicella-zoster virus. (A) Viral DNA and capsid proteins assemble in the nucleus into nucleocapsids which acquire a primary envelope and reach the perinuclear cisterna and rough endoplasmic reticulum (RER). The primary envelope is lost again and naked nucleocapsids reach the cytosol and the *trans*-Golgi network (TGN). (B) Glycosylated envelope lipoproteins are synthetized in the RER. As in other such proteins, their luminal domain is glycosylated by addition of high-mannose sugars to asparagine residues. Tegument proteins are probably synthetized by free ribosomes. The glycoproteins are transported through an internal compartment (IC) to reach the Golgi stack where the original sugars are modified to become complex oligosaccharides. During transport the sideness of glycoproteins is maintained; sugar groups are always found within the lumen of cisternal compartments. Tegument proteins could passively follow the glycoproteins by binding to their cytoplasmic tails or be transported independently. Glycoproteins are finally transported to the TGN, where they accumulate on the concave surface of a TGN-derived sac. Teguments adhere to the cytosolic face of the sac. The concave surface of the sac becomes the definitive viral surface, while the convex surface becomes a transport vesicle, which delivers the viruses to acidic structures identified as prelysosomes. (From Gershon, A.A., Sherman, D.L., Zhu, Z., Gabel, C.A., Ambron, M.T., and Gershon, M., *J. Virol.,* 68, 6372, 1994. With permission.)

HERPESVIRIDAE

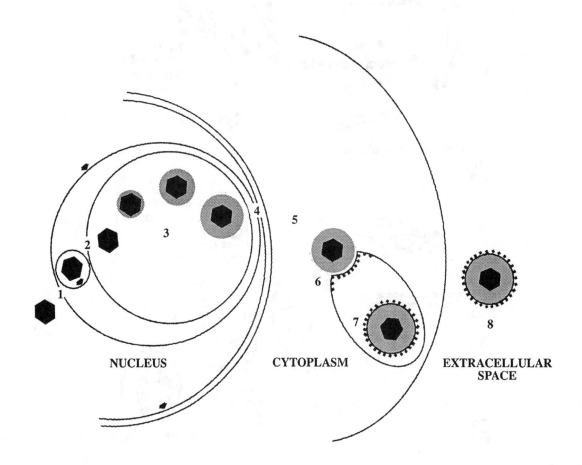

FIGURE 57

Egress of human herpesvirus 6 (HHV-6). Virions have a distinct tegument layer between capsid and envelope. Thymocytes infected with HHV-6 show intranuclear spherical compartments in which tegumentation seems to take place. Termed "tegusomes," these compartments are bounded by two membranes and contain ribosomes. These tegusomes seem to represent a specialized cellular site in which HHV-6 virions acquire their tegument. (From Roffman, E., Albert, J.P., Goff, J.P., and Frenkel, N., *J. Virol.,* 64, 6308, 1990. With permission.)

HERPESVIRIDAE

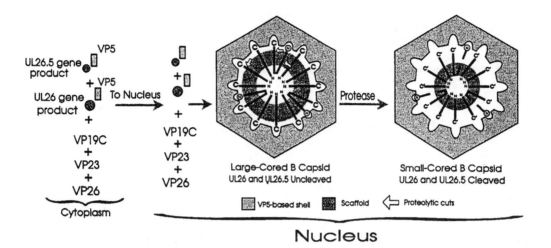

FIGURE 58

Assembly pathway of herpes simplex virus capsid. The major capsid protein VP5 associates in the cytoplasm with precursor proteins UL26 and UL26.5. The complex is transported to the nucleus where capsid proteins assemble into the B capsid which contains a large core of scaffolding protein. Proteolytic cleavage of UL26 and UL26.5 results in a B capsid with a small core that is removed when DNA enters. (From Thomsen, D.R., Newcomb, W.W., Brown, J.C., and Homa, F.L., *J. Virol.,* 69, 3690, 1995. With permission.)

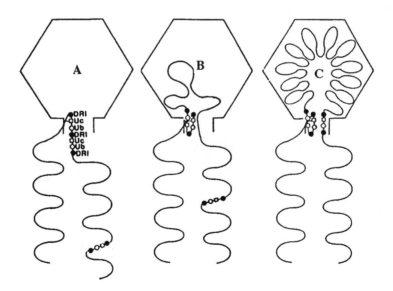

FIGURE 59

DNA packaging in herpes simplex virus. The model requires that proteins attach to the "a" sequence of the DNA concatemer, probably to Uc, and that empty capsids scan concatemeric DNA until contact is made with a protein–Uc complex (A). DNA is then taken in (B) until a "headful" enters the capsid or an "a" sequence in the same orientation as above is encountered (C). (Modified from Roizman, B. and Sears, A.E., in *Virology,* 2nd ed., Vol. 2, Fields, B.N. and Knipe D.M., Eds.-in-chief, Raven Press, New York, 1990, 1795. With permission.)

HERPESVIRIDAE

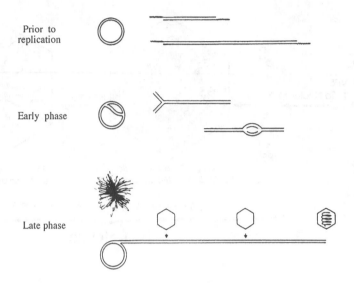

FIGURE 60

Steps in herpesvirus replication. The infecting DNA forms circles. Replication starts mainly on circular but also on linear molecules. Replication of circular DNA is transient. Theta rings, replication forks, and "eye" structures have been observed in the electron microscope. Thereafter, newly synthetized DNA is in the form of concatemers, in which unit-size molecules are found in head-to-tail arrangement. (From Ben-Porat, T., *Curr. Topics Microbiol. Immunol.*, 91, 81, 1981. © Springer-Verlag, Berlin. With permission.)

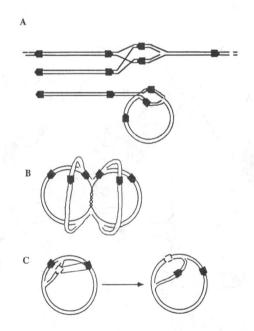

FIGURE 61

Mechanisms of herpesvirus replication which could lead to the formation of large branched structures. Black boxes represent "a" sequences, and arrowheads indicate relative orientation. (A) Formation of branched concatemers by invasion of a terminal "a" sequence. (B) Theta replication of circular genomes, in which partially replicated circles are held together by unreplicated regions. (C) Recombination between a replicated (white) and unreplicated "a" sequence leads to the creation of two replication forks that follow each other. (From Severini, A., Scraba, D.G., and Tyrrell, D.L.J., *J. Virol.*, 70, 3169, 1996. With permission.)

HERPESVIRIDAE

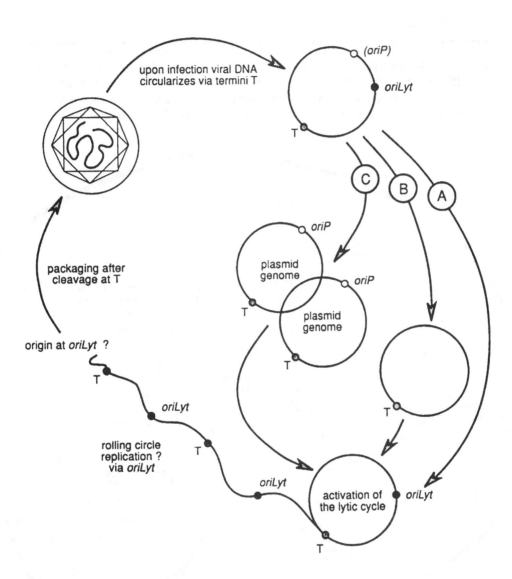

FIGURE 62

Replication and genome maintenance during lytic and latent cycles of different herpesviruses. After infection, DNA ends circularize and form a molecule with ligated termini (T). In "A," the viral genome proceeds to a lytic cycle. DNA replication is initiated at *oriLyt* (also termed *oriL* or *oriS*). DNA replication seems to proceed in a rolling-circle mode and results in a linear concatemer made of multiple head-to-tail linked genome units. This concatemer is cleaved into viral genomes (by a viral endonuclease?) precisely at the site of future termini. The viral DNA is then packaged by a yet unknown mechanism. The lytic cycle results in cell death. In "B," herpesvirus genomes occur as extrachromosomal circular elements in postmitotic elements such as neurons. Virus genomes apparently do not affect proliferation of latently infected cells. Reactivation of viral DNA leads to a lytic cycle. In "C," lymphotropic herpesviruses immortalize or transform host cells by persisting in an episomal state during the latent phase of the viral life cycle. The genome replicates extrachromosomally via *oriP*, the plasmid origin of replication. *oriP* depends on a viral protein for activation of DNA replication. A small proportion of latently infected cells enters into a lytic cycle with or without induction by chemical agents. Lytic DNA replication is initiated by *oriLyt*. (Reprinted from Hammerschmidt, W. and Mankertz, J., *Semin. Virol.,* 2, 257, 1991. By permission of Academic Press Ltd., London.)

5.II.E. IRIDOVIRIDAE

Linear dsDNA
Cubic, naked
Vertebrates, invertebrates

This highly diversified family has four genera and is found in cold-blooded animals (fishes, frogs, insects, worms). Particles are naked icosahedra 125–300 nm in diameter and have multilayered capsids with 3–14% lipids and 812 capsomers. Some members possess an envelope derived from the plasma membrane of the host cell. The replication strategy of iridoviruses is strikingly different from other DNA viruses and involves a nuclear and a cytoplasmic stage.

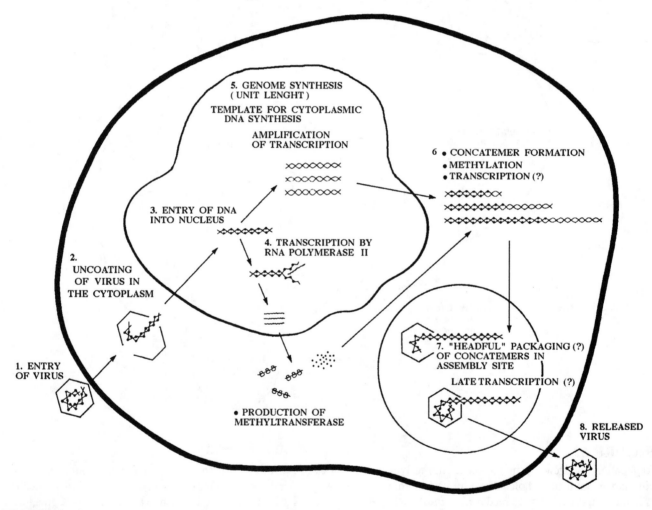

FIGURE 63

Replication of frog virus 3 (FV3). The virus enters cells by pinocytosis, direct penetration, or fusion to the plasma membrane. The genome is uncoated and enters the nucleus where it is transcribed and serves as a template for a first stage of DNA replication. Progeny DNA synthetized in the nucleus serves as a template for further transcriptions or is transported into the cytoplasm for further DNA synthesis. A large concatemer of methylated DNA is produced there and transported to an assembly site. Presumably, progeny DNA is packaged into preformed capsids by a "headful" mechanism, generating circularly permuted and terminally redundant genomes. The whole process resembles that of certain tailed bacteriophages. (Modified from Murti, K.G., Goorha, R., and Granoff, A., *Adv. Virus Res.,* 30, 1, 1985. With permission.)

5.II.F. PAPOVAVIRIDAE

Circular dsDNA
Cubic, naked
Vertebrates

Viruses infect mammals, birds, and reptiles. The family includes the genera *Papillomavirus* (papillomas, warts) and *Polyomavirus* (prototype SV40). Particles are naked icosahedra 40–55 nm in diameter. Papillomaviruses do not replicate in tissue cultures. Viruses enter by endocytosis. DNA replication and particle assembly take place in the nucleus. Replication is first bidirectional and seems to proceed later by a rolling-circle mechanism. Mature viruses are released by lysis. Papovaviruses are able to undergo lytic or temperate cycles and tend to cause tumors or cell transformation *in vitro*. Abortively infected cells may be transformed into a malignant phenotype, a process which may require persistence of at least a part of the viral genome.

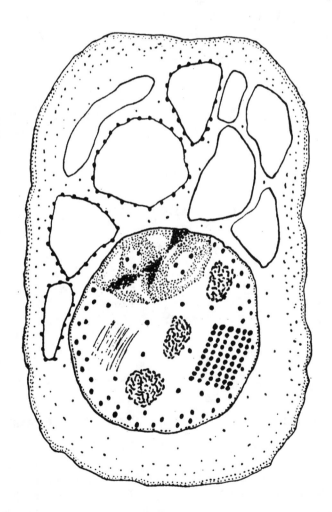

FIGURE 64
Cell showing replication of polyomavirus SV40 ("simian vacuolating agent 40"; authors' note). The nucleus contains virus crystals, bundles of fibrils, and an abnormal nucleolus. The cytoplasm shows intense vacuolization. (From Bernhard, W., in *Ciba Foundation Symposium on Cellular Injury,* De Reuck, A.V.S. and Knight, J. Eds., J. & A. Churchill, London, 1964, 209. With permission.)

PAPOVAVIRIDAE

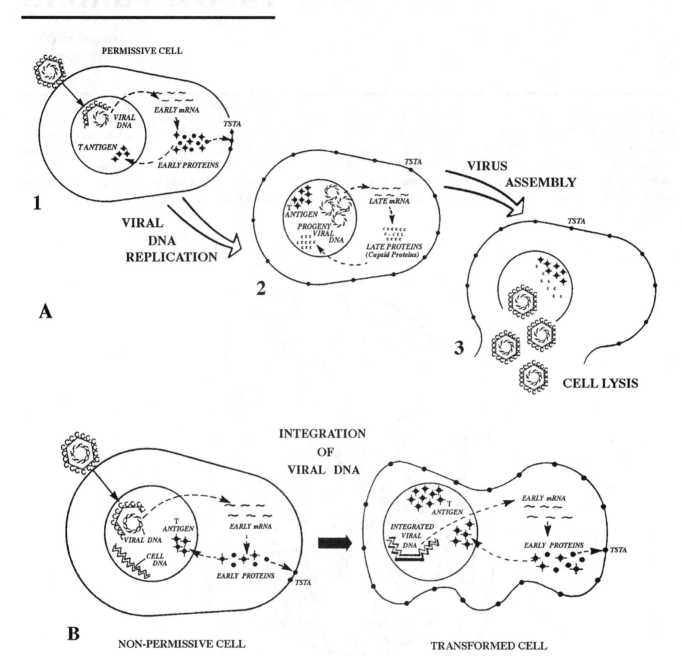

FIGURE 65

(A) Productive papovavirus infection as exemplified by SV40 infection of primary African green monkey kidney cells. (1) During the early phase, early viral proteins are synthetized, including TSTA (tumor-specific transplantation antigen). TSTA is transported to the cell membrane. (2) Late viral DNA is transcribed about 12–15 h after infection. Viral structural proteins are synthetized and transported to the nucleus. (3) Progeny viruses are assembled and accumulate in the nucleus for about 40 h until lysis occurs. (B) Cell transformation as exemplified by SV40 transformation of mouse 3T3 cells. Virus adsorption, penetration, and uncoating occur as in a productive cycle and TSTA is produced, but there is no late phase of replication. Instead, some infected cells integrate the papovavirus genome (or part of it) into their DNA and become permanently transformed. (Adapted from Oxman, M.N., in *Virology and Rickettsiology*, Vol. I, Part 2, Hsiung, G.-D. and Green, R.H., Eds., CRC Press, Boca Raton, FL, 1978, 17. With permission.)

PAPOVAVIRIDAE

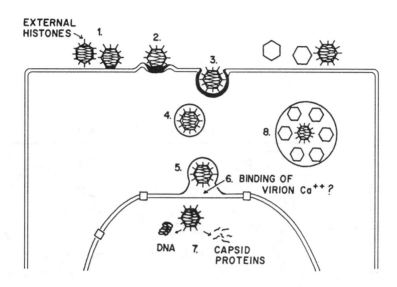

FIGURE 66

Early stages of polyoma virus infection. (1) The virus, carrying a nuclear recognition factor believed to be the virion receptor and consisting of histones, interacts with the host cell receptor. (2) The cell membrane undulates under the virus. (3) The virus penetrates the cell membrane, which closes about the virus particle and pinches it off, forming a monopinocytotic vesicle. (4) The vesicle migrates rapidly to the outer nuclear membrane and (5) fuses with it. (6) The virion is primed for uncoating by chelation of virion-associated calcium(?). (7) The virus enters the nucleus and is uncoated by the action of reducing agents such as cysteine and glutathione. (From Consigli, R.A. and Center, M.S., *CRC Crit. Rev. Microbiol.*, 6, 263, 1978. With permission.)

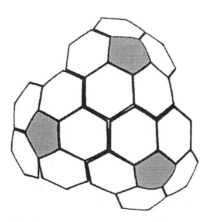

FIGURE 67

Structure of early capsid intermediates in SV40. Potential penton binding sites are shaded. (From Milavetz, B. and Hopkins, T., *J. Virol.*, 43, 830, 1982. With permission.)

PAPOVAVIRIDAE

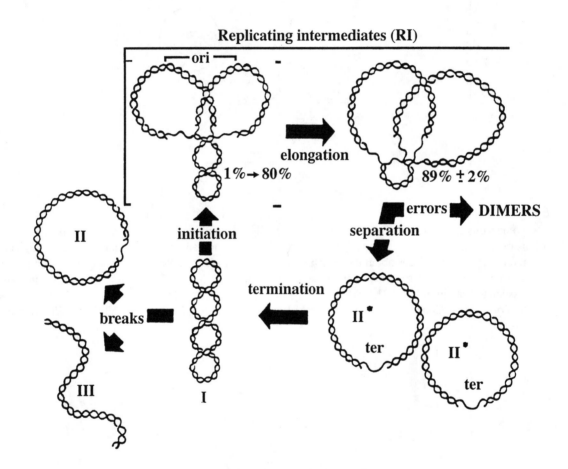

FIGURE 68

DNA replication in SV40. The infecting DNA (I) is covalently closed, superhelical, and circular. Replicating intermediates (RI) contain a superhelical region of unreplicated DNA and two relaxed regions of nonsuperhelical, newly replicated DNA. The resulting II* DNA contains a short gap in the nascent DNA strand at the termination region (*ter*) and is supercoiled again. Random breaks of at least one phosphodiester bond create double-stranded relaxed circles (II) or double-stranded linear DNA molecules one genome in length (III). *ori*, unique origin of DNA replication. (From DePamphilis, M.L. et al., *Cold Spring Harbor Symp. Quant. Biol.*, 43, 679, 1978. With permission.)

5.II.G. POXVIRIDAE

Linear dsDNA
Helical, enveloped
Vertebrates, invertebrates

This very large virus family occurs in a wide variety of mammals and birds (subfamily *Chordopoxvirinae*) and insects (*Entomopoxvirinae*) and includes 11 genera. Particles are brick-shaped or ovoid, measure 220–450 × 140–260 × 110–200 nm, and show considerable morphological variation. They consist of a thin envelope, a lipid layer, lens-shaped lateral bodies, and a core. Chordopoxviruses (eight genera) have two lateral bodies and surface tubules. Entomopoxviruses (three genera) are usually occluded in protein crystals and have one or two lateral bodies and concave or biconcave cores.

Poxvirus infection is characterized by its autonomy from host cell functions. DNA replication, gene expression, and virus assembly take place in the cytoplasm and are mediated by viral enzymes. Viruses enter cells by membrane fusion and endocytosis and are partially uncoated. DNA replication, still incompletely understood, seems to involve a strand displacement mechanism and concatemeric replicative intermediates. Progeny viruses are formed within inclusion bodies or "virus factories." DNA coalesces within lipoprotein bilayers to form spherical particles. Lateral bodies and other constituents are added later. Mature viruses are released by cell lysis or budding from microvilli. Some particles may acquire a second envelope from Golgi membranes or microvilli.

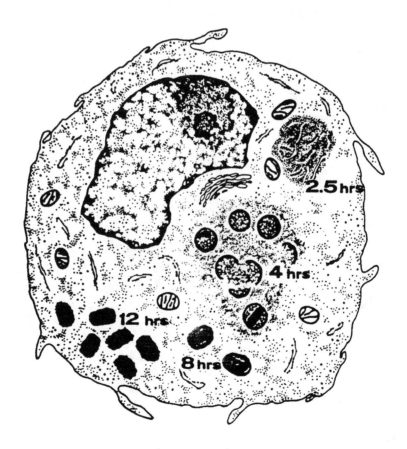

FIGURE 69

A cell at different times after infection with vaccinia virus (drawing by P. Grimley). (From Moss, B., in *Virology*, 2nd ed., Vol. 2, Fields, B.N. and Knipe, D.M., Eds.-in-chief, Raven Press, New York, 1990, 2079. With permission.)

POXVIRIDAE

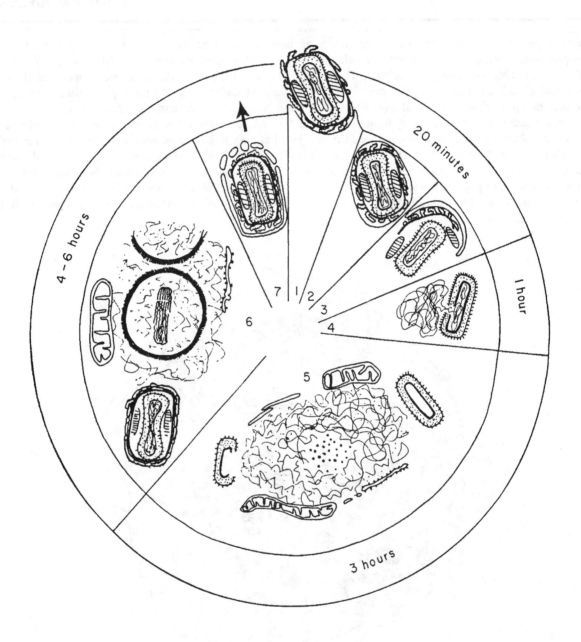

FIGURE 70

Sequence of vaccinia virus uptake and replication in L cells. (Reproduced from Dales, S., *J. Cell Biol.,* 18, 51, 1963. By copyright permission of the Rockefeller University Press.)

POXVIRIDAE

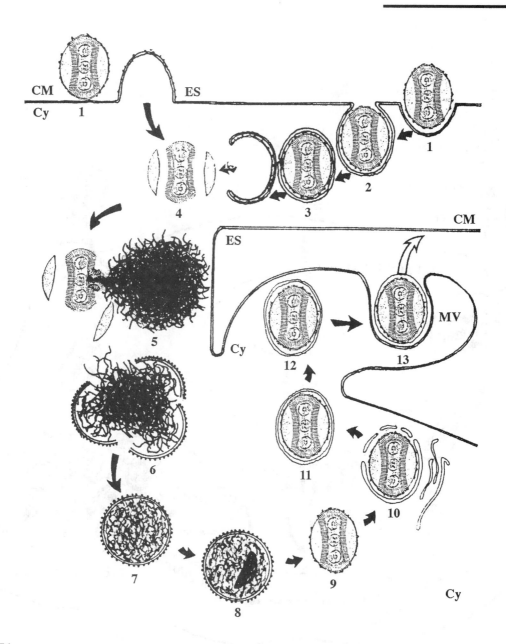

FIGURE 71

Vaccinia virus replication cycle showing sequential stages of viral morphogenesis within the cytoplasm of an infected cell. ES, extracellular space; CM, cell membrane; CY, cytoplasm; MV, microvilli. (1) Virion in contact with cell membrane. (2) Invagination of cell membrane together with attached virion. (3) Fusion of viral envelope with membrane of enclosing vesicle. (4) Release of core and lateral bodies into cytoplasm (this may also result directly from fusion of viral envelope and cell membrane). (5) "Factory" at site of genome release after final uncoating of virus core. (6) Assembly of viral bilayer membrane. (7) Immature virion. (8) Condensation of viral DNA. (9) Mature virion. (10) Virion surrounded by Golgi vesicles or cytoplasmic membrane-like sheath. (11) Virion in cisternae. (12) Migration of the enveloped virion to cell membrane or microvilli. (13) Fusion of cisternal membrane with cell membrane and release of a progeny virion with a single envelope. (By Heinz Hohenberg, Heinrich Pette Institute, Hamburg. From Müller, G. and Williamson, J.D., in *Animal Virus Structure,* Nermut, M.V. and Steven, A.C., Eds., Elsevier, Amsterdam, 1987, 421. With permission.)

POXVIRIDAE

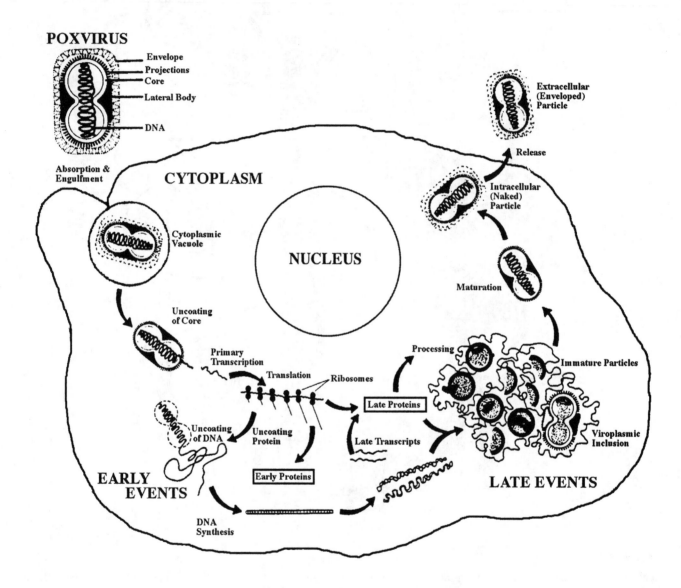

FIGURE 72

The poxvirus replication cycle. (Partially redrawn from Palmer, E.L. and Martin, M.L., *Electron Microscopy in Viral Diagnosis,* CRC Press, Boca Raton, FL, 1988, 175. With permission.)

POXVIRIDAE

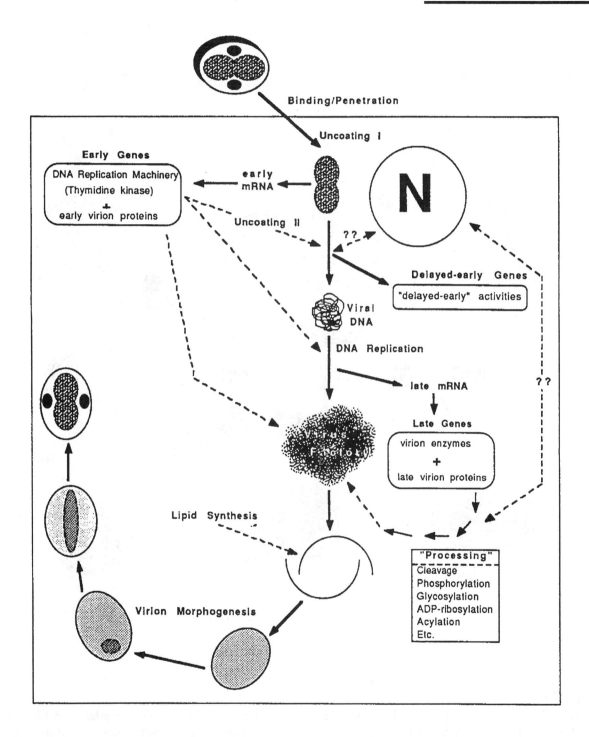

FIGURE 73

Vaccinia virus replication strategy. The events that take place in a host cell that lead to production of mature vaccinia virions. (From VanSlyke, J.K. and Hruby, D.E., *Curr. Topics Microbiol. Immunol.,* 163, 185, 1990. © Springer-Verlag, Berlin. With permission.)

POXVIRIDAE

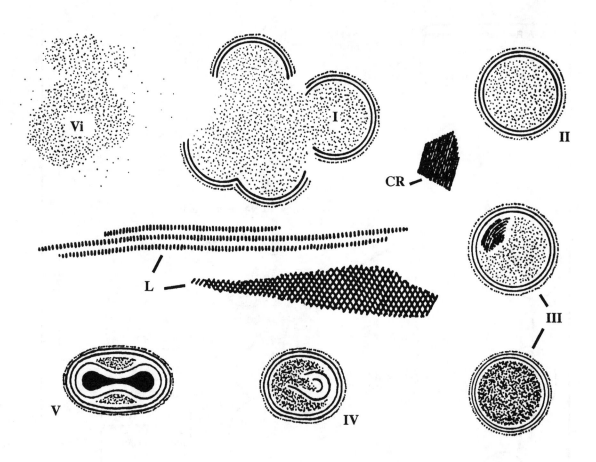

FIGURE 74

Main steps in Shope fibroma virus (genus *Leporipoxvirus*) morphogenesis and illustration of the principal particles seen in inclusion bodies (I–IV). Vi, viroplasm; CR, crystalloid inclusion body; L, lamellae. (From Scherrer, R., *Pathol. Microbiol.*, 31, 129, 1968. Reproduced with permission of S. Karger AG, Basel.)

POXVIRIDAE

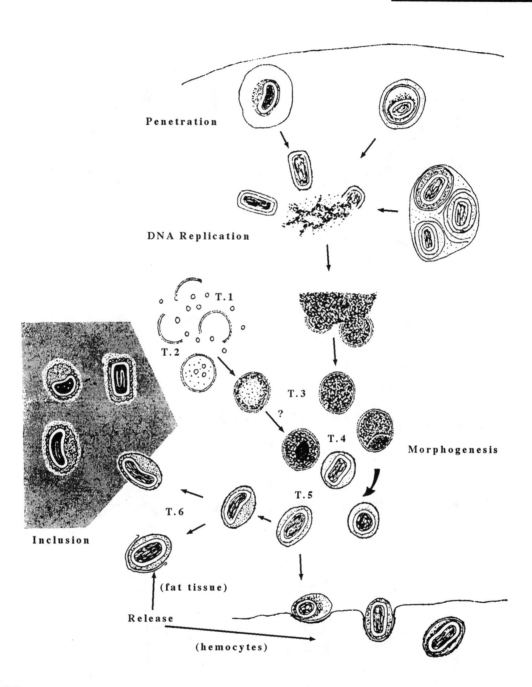

FIGURE 75

Hypothetical replication cycle of *Melolontha melolontha* entomopoxvirus. The principal phases of the cycle were observed in hemocytes, in particular virus penetration, DNA replication, morphogenesis, and release with acquisition of a membrane derived from the cell membrane. Inclusion of particles within proteinic "spherules," very frequent in fat tissue, was exceptional in blood cells. Release with envelope acquisition from the ergastoplasm was observed in fat tissue only. The relationship between empty particles of type 2 and type 3 particles is uncertain. (Adapted from Devauchelle, G., Bergoin, M., and Vago, C., *J. Ultrastruct. Res.,* 37, 301, 1971. With permission.)

POXVIRIDAE

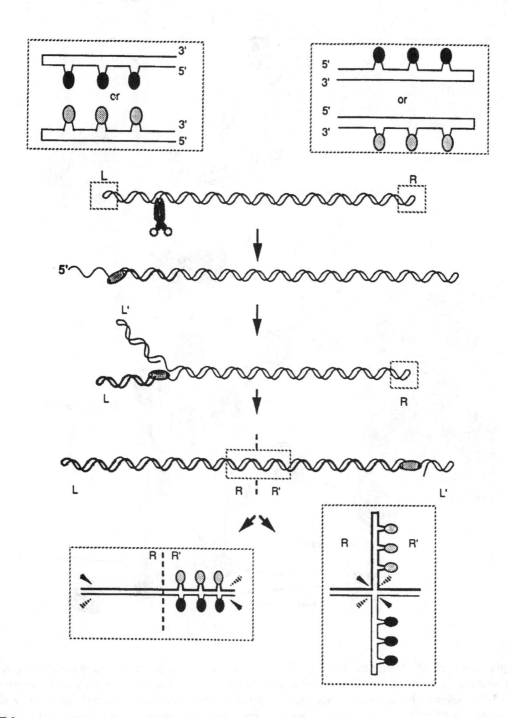

FIGURE 76
Vaccinia virus DNA replication: a working model. A detailed legend appears on the following page.

POXVIRIDAE

LEGEND FOR FIGURE 76

(A) The genome is a linear duplex with covalently closed hairpin termini. Left (L) and right (R) termini are boxed and possible hairpin structures are illustrated above. Hairpins are incompletely base-paired and extrahelical residues are shown as protruding ovals. The two versions of the hairpin are inverted complements of one another. The flip conformation (top) has extrahelical bases in the 5′ half of the sequence; in the flop conformation, the complementary extrahelical bases (shaded) are in the 3′ half of the sequence. (B) Replication presumably starts with introduction of a nick (scissors) near one of the termini. The nick exposes a free 3′OH group which can serve as a primer for the viral DNA polymerase (ellipsoid). The nicked strand is extended by displacement of the looped parental strand (nascent DNA shown in bold). (C) The nascent DNA is a copy of the original terminal loop. Both the parental and nascent strands have the potential to fold back and to form intrastrand duplexes as well as to hybridize with each other. Conversion of the interstrand duplex to two intrastrand duplexes is thought to occur. In this configuration (below), the genome resembles a classical replication fork. The DNA polymerase is now poised to copy one of the parental strands in classical semi-conservation mode. This synthesis can continue rightward around the R hairpin loop (boxed) and copy the previously displaced strand until the (L) terminal loop is reached. (D) This primary product would represent a dimeric intermediate (below). Although higher order concatemers would be formed by subsequent cycles, the dimer is shown here for the sake of simplicity. Vertical dashed line indicates the axis of symmetry of the original R hairpin loop. The dimer contains a duplex copy of the entire hairpin loop. The details of this duplex copy are shown at the bottom left (boxed), with the assumption that the original genome had the flip conformation at the R terminus. If concatemers were processed by simple cleavage and resealing at the junction of R and R′, the genome would have fully base-paired hairpin loops. This model can therefore be ruled out. Instead, the strands must be nicked in a staggered fashion, so that flip and flop conformations can be inherited by progeny genomes. (E) Because the concatemer junction was formed by replication of a self-complementary hairpin loop, each strand can form an intrastrand or an interstrand duplex in the RR′ region. Isomerization of the linear form of the concatemer junction to the cruciform version is shown at bottom right (boxed). Cleavage is thought to occur by introduction of staggered nicks (left) or at the base of the cruciform (right). Two symmetric reactions would occur (black or shadowed arrowheads). Following cleavage directed by the black arrowheads, the flip conformation would be restored at the R terminus, and the flop would be found at the R′ terminus. The R terminus would be derived exclusively from parental sequences, whereas the R′ would be fully derived from nascent sequences. After cleavage directed by shaded arrows, the R terminus would receive a nascent flop hairpin, whereas the R′ terminus would contain a parental flip terminus. Because each concatemeric junction cleavage event should be independent, genomes with all combinations of flip and flop at their termini could be generated. (Reprinted from Traktman, P., *Semin. Virol.*, 2, 291, 1991. By permission of the publisher Academic Press Ltd., London.)

5.II.H. TECTIVIRIDAE

Linear dsDNA
Cubic, naked
Bacteria

This small family is found in bacilli and a few Gram-negative bacteria (enterics, *Thermus*). Particles are icosahedra about 63 nm in diameter and consist of DNA, an outer rigid protein shell, and an inner thick-walled lipoprotein vesicle. Upon adsorption to bacteria or treatment with chloroform, the vesicle transforms itself into a tail-like tube of 60×10 nm and DNA is ejected. DNA replication is protein-primed; its mode is not yet understood. Capsid proteins assemble on elements of the bacterial cytoplasmic membrane to form a procapsid which is filled with DNA. Phages are released by cell lysis.

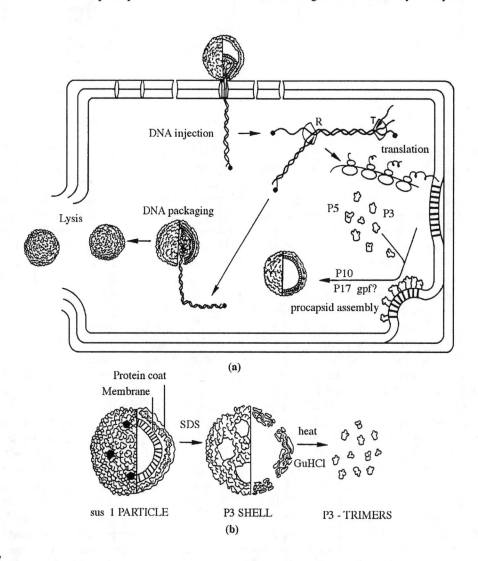

FIGURE 77

Life cycle and structure of PRD1. (a) Life cycle. (b) Morphology of subviral particles. The *sus*1 particle is produced in large quantities by a mutant viral strain. Minor coat protein subunits (P5) are indicated by black pentagons. Treatment with SDS eliminates the viral membrane and coat pentons along with peripentonal P3 subunits, generating a shell of 180 P3 trimers. The major coat protein P3 constitutes 70–80% of the total protein mass of the native virion. (Modified from Tuma, R., Bamford, J.K.H., Bamford, D.H., Russell, M.P., and Thomas, G.J., *J. Mol. Biol.*, 257, 87, 1996. With permission.)

TECTIVIRIDAE

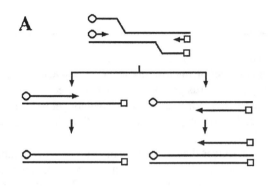

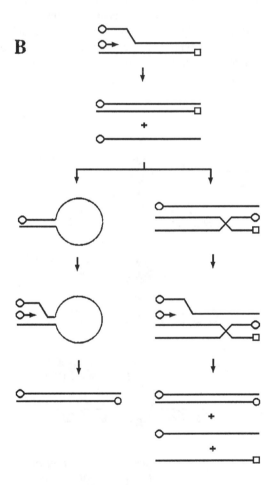

FIGURE 78

Possible DNA replication pathways of PRD1. (A) In the replicative pathway, replication starts at both DNA ends. Separation of the two daughter molecules occurs when both replication forks meet. This pathway explains the observation that phages with different ITRs at DNA ends are viable. (B) The sequence-monitoring pathway implies that replication starts at one DNA end only and allows for the recovery of ITR sequences. Recovery can occur by either formation of a panhandle (left) or by an intramolecular recombination event (right). These processes involve only DNA sequences corresponding to ITRs (110 bp). Circles and squares indicate differences in ITR sequences. (From Bamford, D.H., Caldentey, J., and Bamford, J.K.H., *Adv. Virus Res.*, 45, 281, 1995. With permission.)

III. dsDNA Viruses with Binary Symmetry

Linear dsDNA
Binary, naked

Viruses of this type consist of heads and tails of cubic and helical symmetry, respectively, and are commonly called "tailed phages." With about 4500 isolates studied by electron microscopy and a host range that includes archae-, eu-, and cyanobacteria,[81] tailed phages are the largest and probably oldest virus group. They are classified into three families (*Myoviridae, Siphoviridae, Podoviridae*) and 14 genera; however, the vast majority of isolates have not yet been attributed to any genus and more genera are likely to be defined in the near future. Tailed phages are extremely diversified with respect to particle size, fine structure, and physicochemical and biological properties.

Capsids are icosahedra or elongated derivatives thereof, usually measure 60 nm in diameter (range 45–170 nm), and contain one molecule of DNA. Elongated phage heads, up to 230 nm in length, are more rare than isometric capsids. Tails consist of helical rows or stacked disks of subunits. Phages are virulent or temperate. They adsorb to bacterial surfaces by tail tips, digest the cell wall locally, and inject their DNA. In the lytic (productive) cycle, DNA replication is semi-conservative and usually results in the formation of concatemers. Particles assemble via separate pathways for heads and tails. The head pathway comprises one or more procapsids which are minimally composed of scaffolding, capsid, and portal vertex protein. DNA is cut to size and enters procapsids which mature by proteolytic cleavage. Progeny phages are liberated by cell lysis. In the temperate or lysogenic cycle, the infecting DNA integrates into the host genome or persists in the cytoplasm as a plasmid. It remains there in a latent state until the "prophage" is set free, either spontaneously or by induction, and enters a regular lytic cycle.

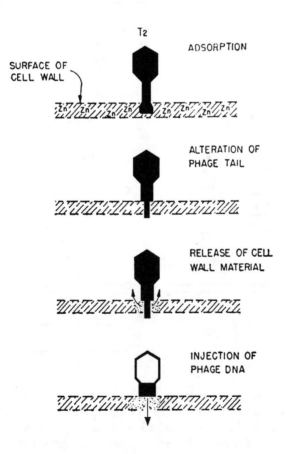

FIGURE 79

First representation of infection by bacteriophage T2. (From Evans, E.A., *Fed. Proc.*, 15, 827, 1956. With permission.)

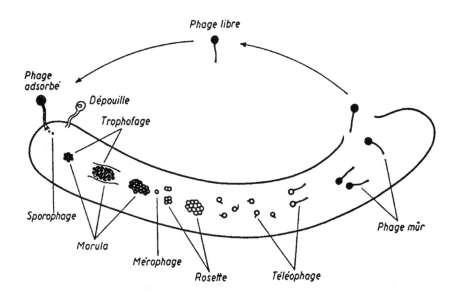

FIGURE 80

First representation of a bacteriophage life cycle, starting with infection at left and ending with liberation of "ripe" phages (*phage mûr*) at right. (Early observations suggested that phages underwent development cycles. It was thought that the cycle started with some kind of insemination and a special vocabulary, derived from embryology and cytology, was created; authors' comment.) (From Penso, G., in *Problèmes Actuels de Virologie,* Hauduroy, P., Ed., Masson, Paris, 1954, 107. With permission.)

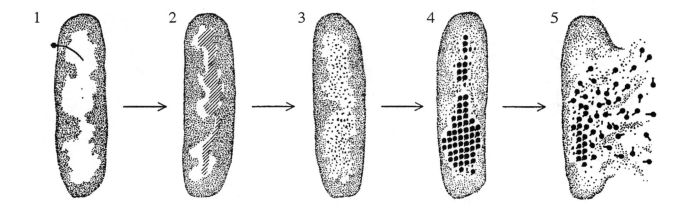

FIGURE 81

View of the lytic cycle. (1) Infection begins when viral DNA enters the bacterium. (2) Bacterial DNA is disrupted and viral DNA is replicated. (3) Synthesis of viral proteins and (4) their assembly continues until (5) the cell bursts, releasing particles. (From Wood, W.B. and Edgar, R.S., *Sci. Amer.,* 217, 61, 1967. With permission.)

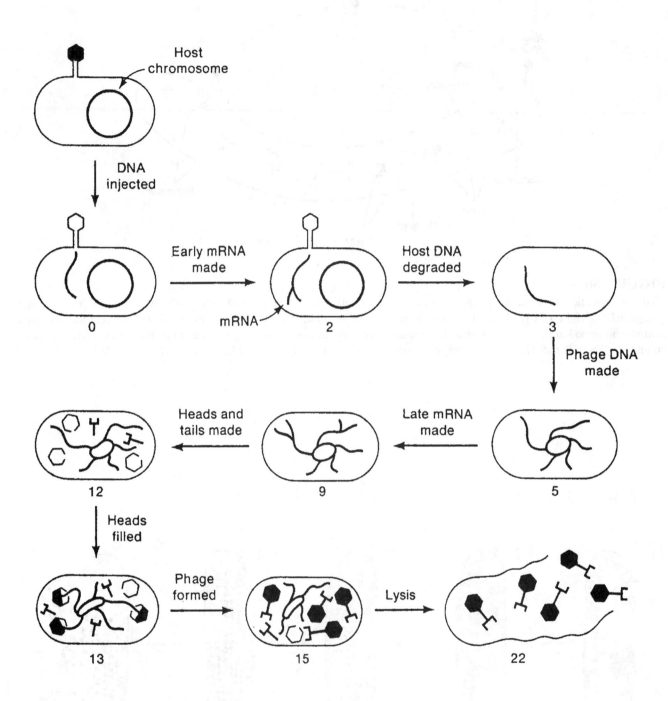

FIGURE 82

The lytic cycle as exemplified by phage T4. Numbers represent time after injection in minutes. For clarity, mRNA is drawn only at the time at which its synthesis begins. (From Freifelder, D., *Molecular Biology, A Comprehensive Introduction to Prokaryotes and Eukaryotes,* Science Books International, Boston, 1983, 616. © 1983 Boston: Jones and Bartlett Publishers. Reprinted with permission.)

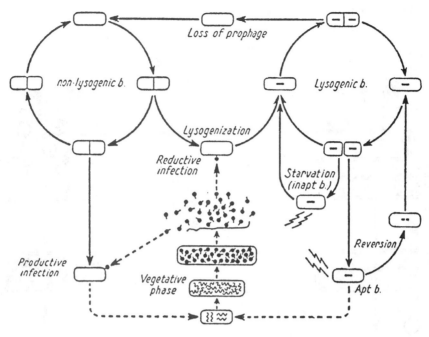

FIGURE 83

First representation of the temperate cycle. Infection of nonlysogenic bacteria results in phage production or lysogenization by phage genomes (prophages). Lysogenic bacteria may carry prophages indefinitely, lose them, or be induced by certain agents, symbolized by lightning rods, to produce new phages. (The now obsolete term "gonophage" designates a replicating phage genome; authors' comment.) (From Lwoff, A., *Bacteriol. Rev.*, 17, 269, 1953. With permission.)

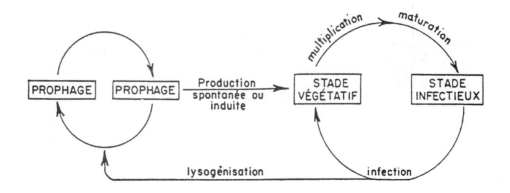

FIGURE 84

Another early representation of the temperate cycle. (From Jacob, F., *Les Bactéries Lysogènes et la Notion de Provirus*, Masson, Paris, 1954, 30. With permission.)

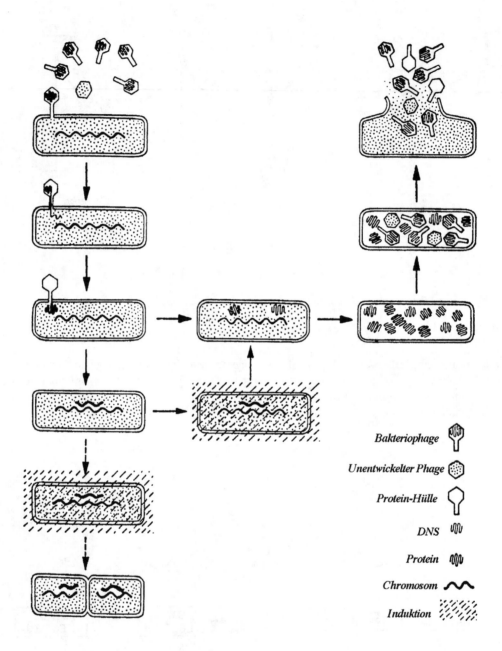

FIGURE 85

Early representation of lytic and temperate cycles. After adsorption of the phage, its DNA content enters the bacterium (above left). The DNA is replicated (center) and new phages are produced with concomitant destruction of the bacterium (above right). Alternatively, the phage DNA becomes a prophage which either is resistant to inducing agents (e.g., UV or X-rays) or is made by these agents to replicate and to produce new phages (center and above right). (The prophage is shown as a plasmid and not as an integrated entity; authors' comment.) (From Hercík, F., *Biophysik der Bakteriophagen,* VEB Deutscher Verlag der Wissenschaften, Berlin, 1959, 5. With permission of Hüthig GmbH, Heidelberg.)

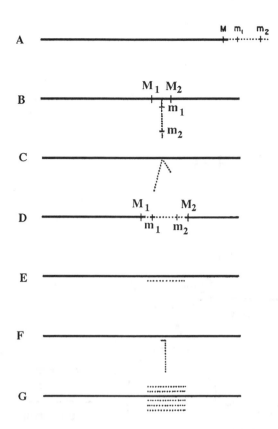

FIGURE 86

Early models for prophage attachment to the bacterial chromosome. Model D anticipates the Campbell model (below). M_1 and M_2, bacterial markers; m_1 and m_2, prophage markers. Solid line, bacterial chromosome; dotted line, prophage. (From Bertani, G., *Adv. Virus Res.,* 5, 151, 1958. With permission.)

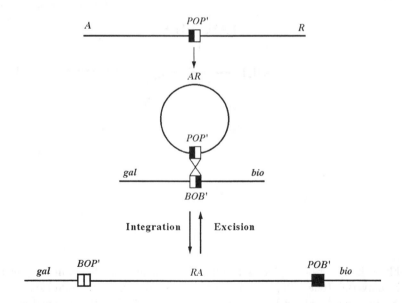

FIGURE 87

The Campbell model[88] for prophage λ integration and excision. The phage attachment site has been named POP′ in accordance with subsequent findings. The bacterial attachment site is BOB′. The integrated prophage is flanked by two new attachment sites named BOP′ and POB′. (From Freifelder, D., *Molecular Biology, A Comprehensive Introduction to Prokaryotes and Eukaryotes,* Science Books International, Boston, 1983, 672. © 1983 Boston: Jones and Bartlett Publishers. Reprinted with permission.)

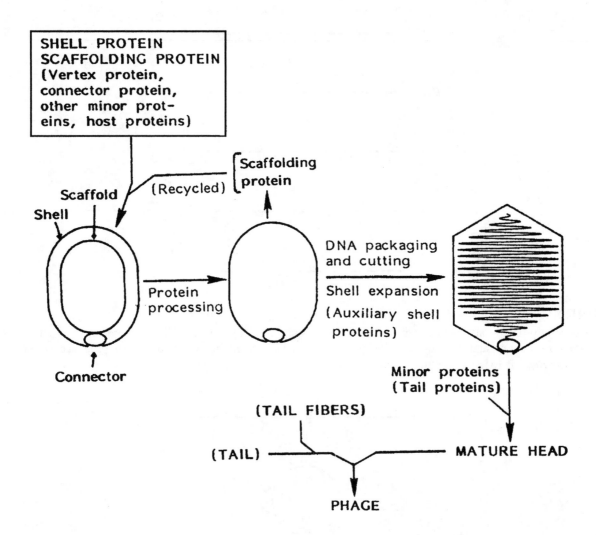

FIGURE 88

Generalized head assembly pathway for tailed phages. Features in parentheses are found in some phages but not in others; all other features appear to be universal. Head and proheads are shown as being prolate. (From Hendrix, R.W., in *Virus Structure and Assembly,* Casjens, S., Ed., Jones and Bartlett Publishers, Boston, © 1985, 169. Reprinted with permission.)

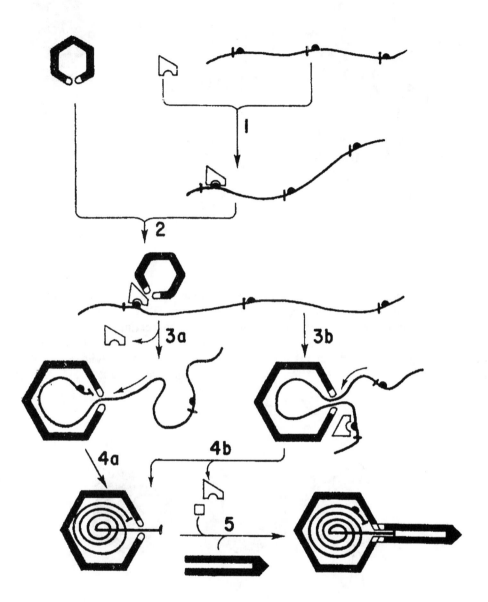

FIGURE 89

DNA packaging in tailed phages. (1) A specific base sequence of phage DNA is bound by a recognition protein, which is usually a multimer of two phage proteins. (2) Proheads bind to the protein–DNA complex. (3 and 4) DNA enters the prohead, with the first DNA portions occupying positions at greater radius. These steps are complex, including prohead expansion and cutting of packaged DNA from the concatemer. Evidence from the study of P22 indicates that (4b) the recognition protein leaves only after the DNA is fully packaged and cut from the concatemer. (5) Stabilizing proteins and tails are added. In several phages, for example λ, the DNA is known to protrude into the central channel of the tail. (From Casjens, S., in *Virus Structure and Assembly,* Casjens, S., Ed., Jones and Bartlett Publishers, Boston, © 1985, 75. Reprinted with permission.)

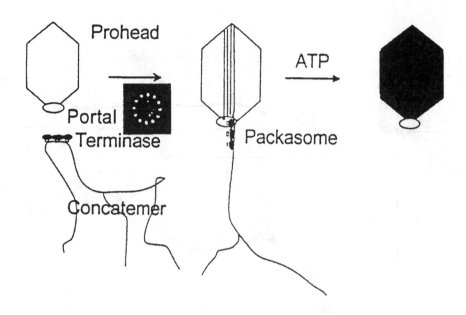

FIGURE 90

Cutting and packaging of concatemeric DNA into proheads. After formation of a prohead, terminases (generally comprising a small and a large subunit) produce a cut in linear or branched concatemeric DNA and introduce the DNA through the portal vertex of the prohead, shown head-on with 12-fold symmetry and a central channel (inset). Terminase and portal vertex interact closely and may constitute a head-filling "packasome," which translocates DNA by hydrolyzing ATP. Expansion of the prohead shell occurs during packaging, and the concatemer is again cleaved when a headful of DNA is packaged. Finally, a tail is joined to the head. (From Black, L.W., *BioEssays*, 17, 1025, 1995. With permission of ICSU Press, Cambridge, UK.)

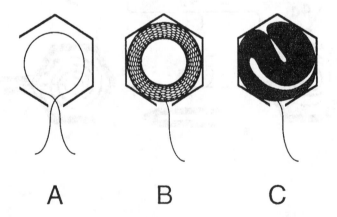

FIGURE 91

Model for formation of DNA toroids within bacteriophage heads. (A) The first DNA to enter the head forms an initial loop. (B) Successive loops are deposited on the inside. DNA condenses as it enters the phage head, causing gyration of the toroid so that the point of DNA deposition is always near the proximal vertex. (C) The toroid collapses upon itself when it has grown to a point that it becomes constrained by the capsid. Head filling continues until either the folded toroid is no longer able to gyrate within the capsid or the entering DNA is cut by a nuclease. (From Hud, N.V., *Biophys. J.*, 69, 1355, 1995. With permission.)

5.III.A. MYOVIRIDAE

Contractile tails

The *Myoviridae* family presently includes six genera and about 1200 members. Heads are isometric or elongated. Tails measure 80–455 × 16–20 nm and consist of a central tube and a contractile sheath separated from the head by a neck. Contraction of the sheath brings the tube in contact with the bacterial cytoplasmic membrane and starts infection. Myoviruses tend to have larger heads and higher particle weights and DNA contents than other tailed phages. Although the family is exemplified by bacteriophage T4, the properties of this particular virus are not necessarily those of other myoviruses. The T4 genome is circularly permuted and terminally redundant (Fig. 91). Other well-known myoviruses illustrated by diagrams are coliphages Mu, P1, P2, and P4. The latter phage is defective and requires the presence of helper virus P2 for all morphogenetic and lytic functions.

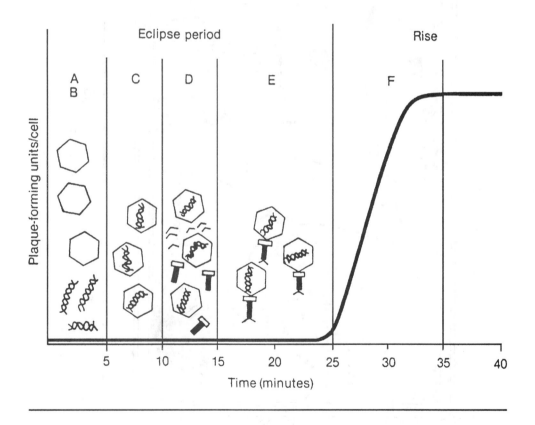

FIGURE 92

Growth curve of bacteriophage T4. Intervals A–F depict approximate time for production and assembly of viral components. Mature viral particles are not assembled until 20–25 min following infection. Sudden rise in detectable viral particles is due to lysis of bacterial cells and immediate release of all progeny viral particles. Much longer growth periods are observed in tailed phages of halobacteria (up to 17 h) and cyanobacteria (up to 40 h). The shape of phage heads has been misrepresented (authors' comments). (From Boyd, R.F., *General Microbiology,* Times Mirror/Mosby College, St. Louis, 1984, 358. Reproduced with permission of The McGraw-Hill Companies.)

MYOVIRIDAE

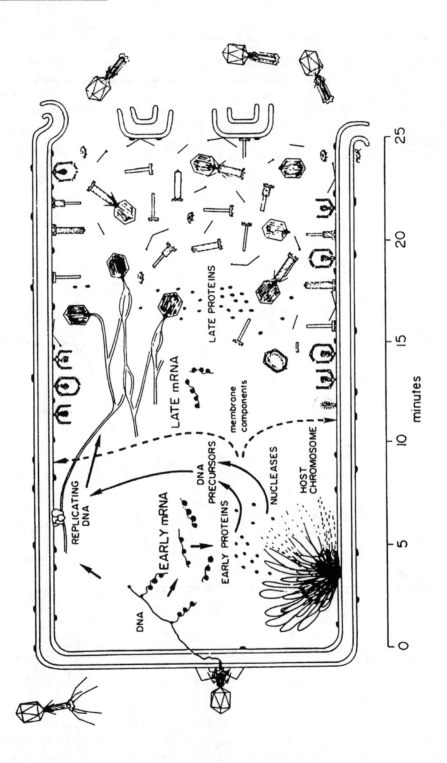

FIGURE 93
Overview of the T4 development program (drawing by F.A. Eiserling). (From Mathews, C.K., in *Molecular Biology of Bacteriophage T4,* Karam, J.D., Ed.-in-chief, American Society for Microbiology, Washington, D.C., 1994, 1. With permission.)

MYOVIRIDAE

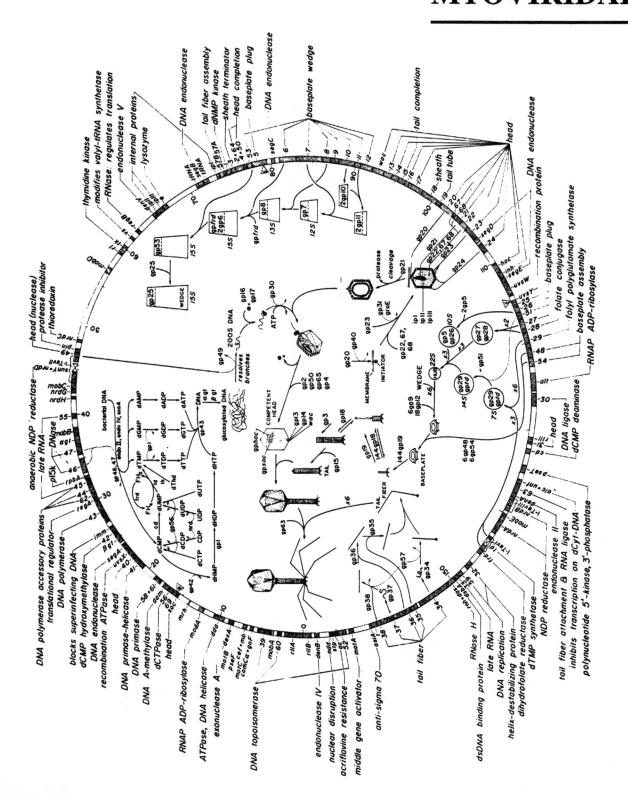

FIGURE 94

Bacteriophage T4 synthesis as related to the T4 genetic map (drawing by E. Kutter). (From Karam, J.D., Ed.-in-chief, *Molecular Biology of Bacteriophage T4,* American Society for Microbiology, Washington, D.C., 1994, ii. With permission.)

MYOVIRIDAE

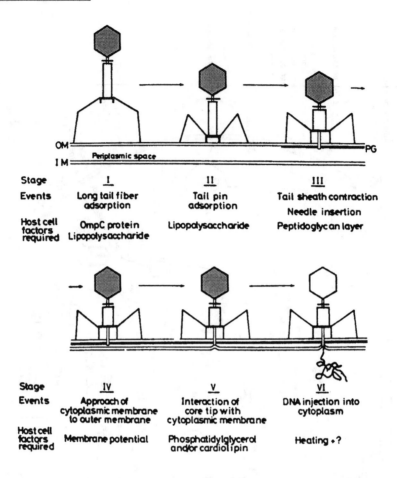

FIGURE 95

Infection by bacteriophage T4. OM, outer membrane; CM, cytoplasmic membrane; PG, peptidoglycan layer. (From Furukawa, H., Kuroiwa, T., and Mizushima, S., *J. Bacteriol.,* 154, 938, 1983. With permission.)

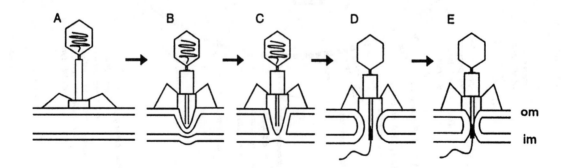

FIGURE 96

Hypothetical induction of membrane fusions by T4. (A) Adsorption. (B) Invagination of outer membrane (om) to reach the inner membrane (im). (C) Membrane fusion. (D) Injection of DNA, phase of K⁺ efflux as a result of leakiness. (E) Sealing reaction between membranes and phage tail. (From Dreiseikelmann, B., *Microbiol. Rev.,* 58, 293, 1994. With permission.)

MYOVIRIDAE

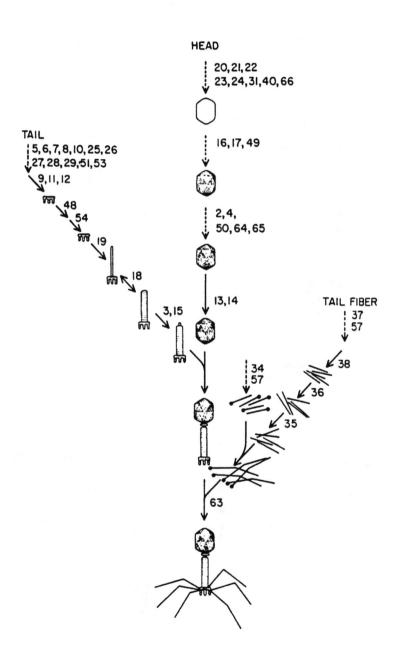

FIGURE 97

Partial sequence of steps in T4 morphogenesis. Numbers indicate T4 genes involved. (From Wood, W.B., Dickson, R.C., Bishop, R.J., and Revel, H.R., in *Generation of Subcellular Structure,* First John Innes Symposium, Norwich, 1973, Markham, R., Ed., Elsevier, Amsterdam, 1973, 25. With permission.)

MYOVIRIDAE

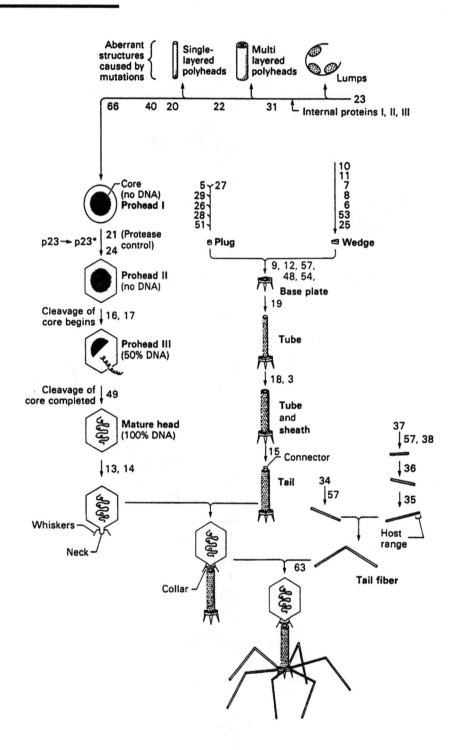

FIGURE 98

A complete view of T4 assembly. Head, tail, and tail fibers are assembled separately. Numbers indicate T4 genes involved; p23* is a cleavage product of protein 23. Aberrant head structures accumulate when a mutation prevents the function of the next gene in the assembly line. "Lumps" are disorganized masses of p23 at the plasma membrane. (From Dulbecco, R. and Ginsberg, H.S., in *Microbiology,* 3rd ed., Davis, B.D., Dulbecco, R., Eisen, H., and Ginsberg, H.S., Eds., Harper & Row, Hagerstown, MD, 1980, 885. With permission.)

MYOVIRIDAE

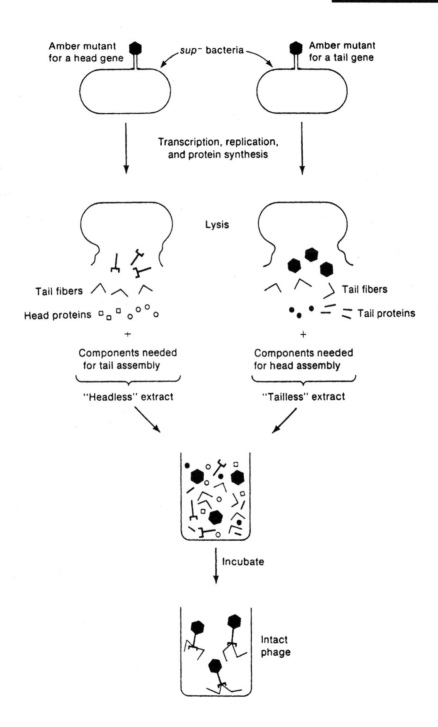

FIGURE 99

Complementation assay for reconstitution of the T4 assembly pathway. If two extracts of infected cells, one lacking heads and the other lacking tails, are mixed, complete phage particles form *in vitro*. This approach can be extended to individual proteins. The assay requires a large collection of mutants unable to produce complete particles. (From Freifelder, D., *Molecular Biology, A Comprehensive Introduction to Prokaryotes and Eukaryotes,* Science Books International, Boston, 1983, 626. © 1983 Boston: Jones and Bartlett Publishers. Reprinted with permission.)

MYOVIRIDAE

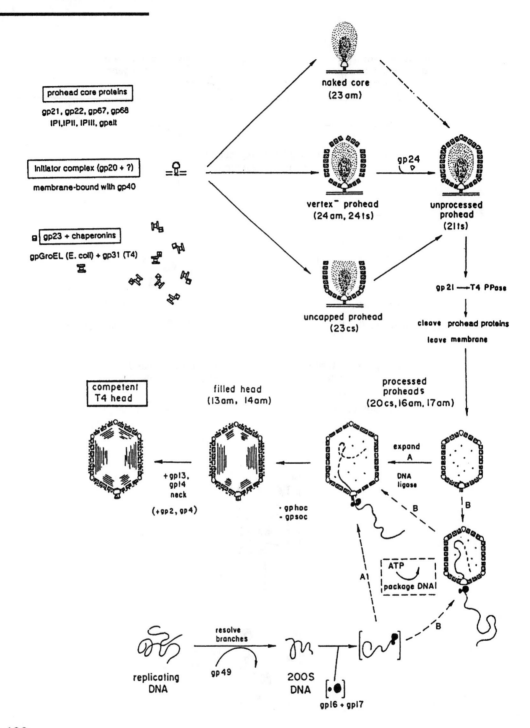

FIGURE 100

T4 head assembly pathway. Successive phases include (1) synthesis of precursor proteins, notably of gp23, precursor of the major capsid protein; (2) assembly of the unprocessed prohead; and (3) prohead processing and maturation. Concatemeric DNA has a separate maturation pathway which converges with that of the prohead. There are two possible modes for the initiation of DNA packaging, involving (A) expanded and (B) unexpanded proheads. (From Black, L.M., Showe, M.K., and Steven, A.C., in *Molecular Biology of Bacteriophage T4,* Karam, J.D., Ed.-in-chief, American Society for Microbiology, Washington, D.C., 1994, 218. With permission.)

MYOVIRIDAE

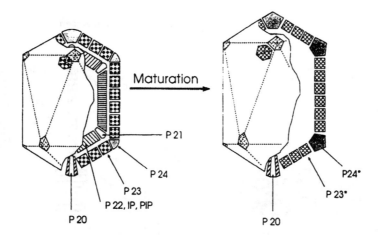

FIGURE 101

Models for structure and composition of the T4 prohead and mature head. Prohead formation may be initiated by assembly of p23 (its major component) at a p20-containing structure with fivefold symmetry at the tail attachment site. p22 is the principal, and IP and PIP are minor constituents of the prehead core. p24 is the vertex protein which controls the activation of the protease from its p21 precursor. Maturation is the result of limited proteolysis of p23 and p24 and complete degradation of p21, p22, IP, and PIP. (From Hellen, C.U.T. and Wimmer, E., *Experientia,* 48, 201, 1992. With permission.)

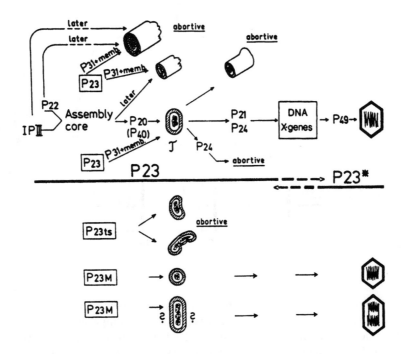

FIGURE 102

T4 head assembly. Many side tracks lead to abortive polymorphic variants. (From Aebi, U. et al., *J. Supramol. Struct.,* 2, 253, 1974. © 1974 John Wiley & Sons. Reprinted by permission of Wiley-Liss, Inc., a subsidiary of John Wiley & Sons, Inc.)

MYOVIRIDAE

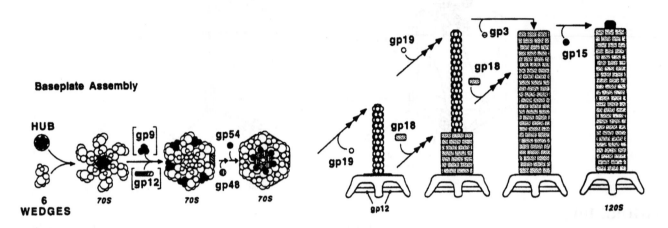

FIGURE 103

T4 base plate and tail assembly. A top view of the base plate shows polymerization of wedges and hub with positions of gp9, gp12, gp48, and gp54. (Modified from Coombs, D.H. and Arisaka, F., in *Molecular Biology of Bacteriophage T4,* Karam, J.D., Ed.-in-chief, American Society for Microbiology, Washington, D.C., 1994, 259. With permission).

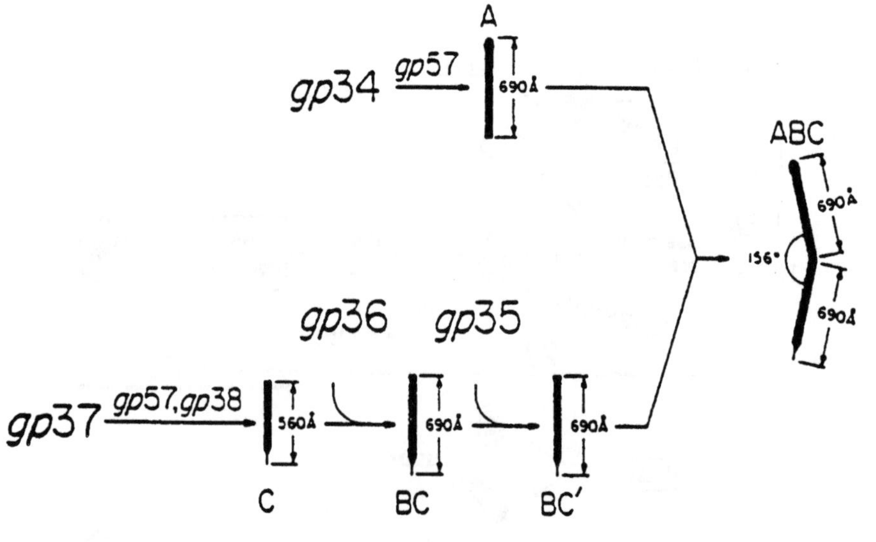

FIGURE 104

T4 tail fiber assembly. Tail fibers consist of four different proteins, constituting the proximal half fiber (gp 34) and distal half fiber (gp35–37), respectively, and corresponding to three antigens (A–C). The completed distal half fiber is designated BC′ to indicate that it is structurally different but serologically indistinguishable from its precursor BC (10 Å = 1 nm). gp38 and gp57 are nonstructural catalysts. (Modified from Bishop, R.J. and Wood, W.B., *Virology,* 72, 244, 1976. With permission.)

MYOVIRIDAE

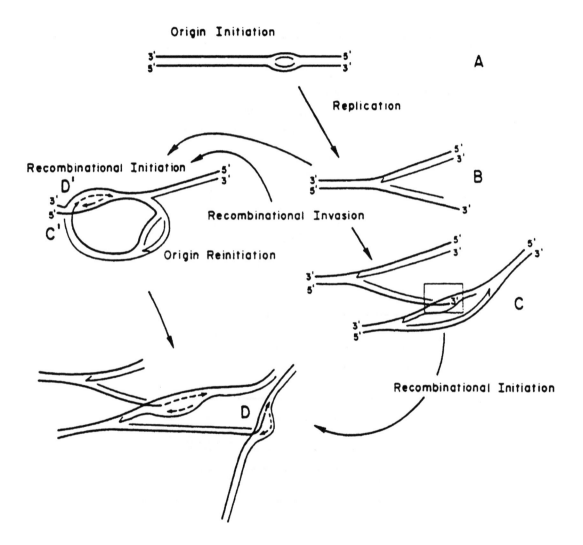

FIGURE 105

T4 DNA replication. Heavy solid lines indicate parental DNA. (A) Initiation of replication at an origin sequence. (B) When a growing point reaches one end of the chromosome, the tip of the template for the lagging strand remains single-stranded. This single-stranded segment invades a homologous segment of another chromosome (C) or of the same molecule (C′) to give a recombinational fork. This fork can be cut and joined to give covalently linked recombinants (D, D′) and DNA fragments. (Modified from Dannenberg, R. and Mosig, G., *J. Virol.,* 45, 813, 1983. With permission.)

MYOVIRIDAE

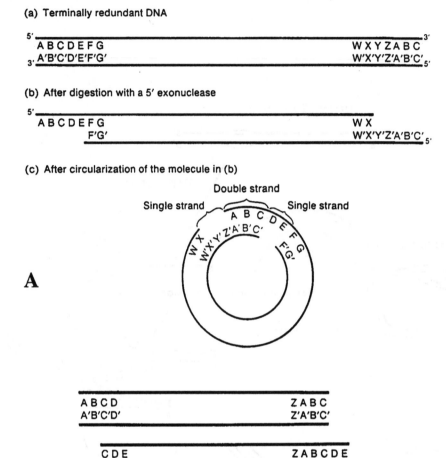

(a) Terminally redundant DNA

5' _____ 3'
   A B C D E F G   W X Y Z A B C
3' A'B'C'D'E'F'G'   W'X'Y'Z'A'B'C' 5'

(b) After digestion with a 5' exonuclease

5' _____
   A B C D E F G   W X
             F'G'   W'X'Y'Z'A'B'C' 5'

(c) After circularization of the molecule in (b)

Double strand

Single strand   A B C D   Single strand

                    W X   E F G
               W'X'Y'Z'A'B'C'   F'G'

A

B

A B C D   Z A B C
A'B'C'D'   Z'A'B'C'

C D E   Z A B C D E
C'D'E'   Z'A'B'C'D'E'

E F G   Z A B C D E F G
E'F'G'   Z'A'B'C'D'E'F'G

G H I   Z A B C D E F G H I
G'H'I'   Z'A'B'C'D'E'F'G'H'I'

FIGURE 106

Consequences of T4 DNA encapsidation. T4 DNA (as that of many other tailed phages) becomes terminally redundant and circularly permuted through encapsidation. (A) Terminal redundancies occur in phages whose DNA is not cut at specific sites and whose heads are filled to capacity ("headful hypothesis"). If the quantity of phage DNA accommodated by a phage head is superior to the normal genome length, progeny DNA is cut at variable sites and may thus contain identical genes in a variable order (permutation). This type of DNA gives circular molecules after exonuclease digestion or denaturation and subsequent renaturation (center). (B) A whole population of terminally redundant DNA molecules may be produced in this way. (From Freifelder, D., *Molecular Biology, A Comprehensive Introduction to Prokaryotes and Eukaryotes,* Science Books International, Boston, 1983, 612. © 1983 Boston: Jones and Bartlett Publishers. Reprinted with permission.)

MYOVIRIDAE

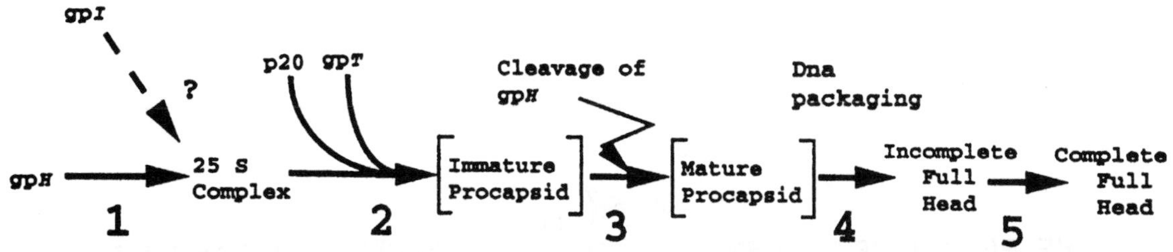

FIGURE 107

Assembly of Mu procapsids. (1) A 25S initiator complex is formed. (2) gpT polymerizes on the initiator and forms a head shell. This would require scaffolding protein gp20 and would generate immature procapsids containing gpT, the uncleaved form of portal protein gpH, and scaffolding protein gp20. (3) gpH is cleaved and the scaffolding protein is removed, yielding a mature procapsid able to accept DNA. (4) Phage DNA is packaged. (5) Head completion. (From Grimaud, R., *Virology,* 217, 200, 1996. With permission.)

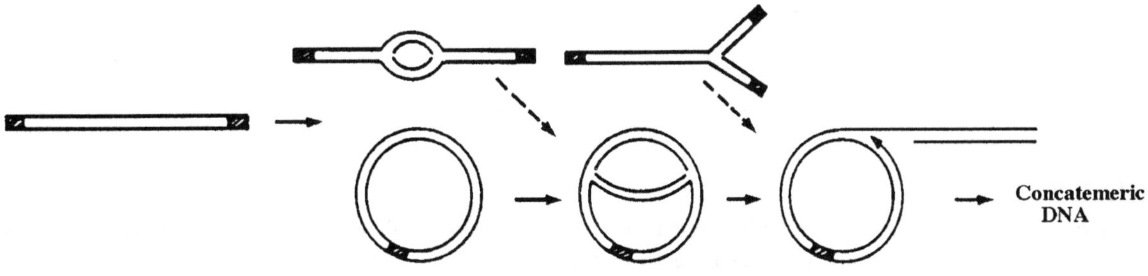

FIGURE 108

Replication of P1 DNA. A P1 site-specific recombination system controls the circularization of phage DNA (left). A later, probably rolling-circle mode is controlled by the bacterial general recombination system, Rec, and is presumed to lead to the formation of concatemers (right). Shaded sites indicate the sites *loxP,* at which the P1 recombination system acts. (From Cohen, G., *Virology,* 131, 159, 1983. With permission.)

MYOVIRIDAE

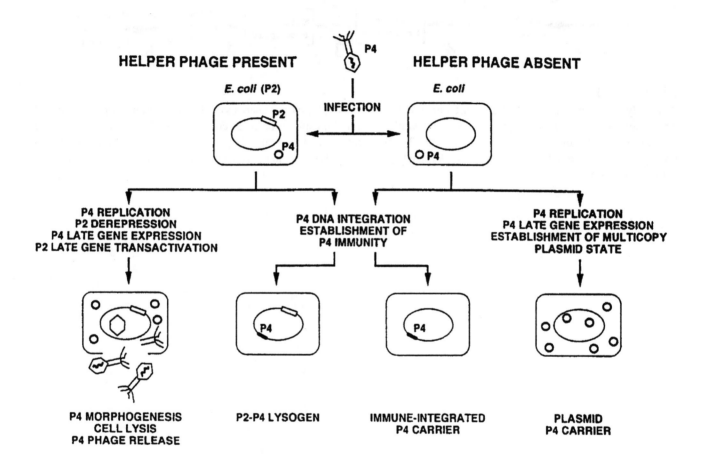

FIGURE 109

The P4 life cycle. When P4 infects *E. coli* harboring a genome of P2, it enters either the lysogenic or the lytic pathway. In the absence of a helper phage, infection by P4 leads to an integrated condition or the establishment of a multi-copy state of maintenance. (From Dehò, G., Bertani, G., and Polissi, A., in *Pseudomonas, Molecular Biology and Biotechnology,* Galli, E., Silver, S., and Witholt, B., Eds., American Society for Microbiology, Washington, D.C., 1992, 358. With permission.)

MYOVIRIDAE

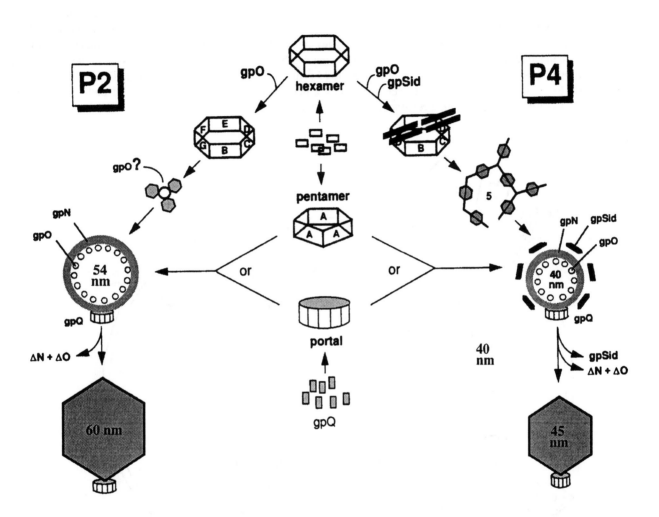

FIGURE 110

P2 and P4 head assembly. Although P2 and P4 build their capsids from the same precursor, their heads differ in size. P2 builds a 60-nm, T=7 capsid from 420 subunits, whereas P4 makes a 45-nm, T=4 capsid from 240 subunits. This difference leads to substantial changes in shell geometry. In the present model, gpN precursors (horizontal open rectangles) are assumed to form hexamers, which associate with scaffolding proteins gpSid (filled bars) and gpO (open circles). The letters B to G indicate the quasi-equivalent position of gpN in hexameric capsomers. In P4, gpSid-gpN oligomers become connected by Sid bridges into pentameric rings with trivalent branching at every threefold axis. This produces a dodecahedral outer cage containing a gpN shell about 40 nm wide. In P2, a speculative function of gpO (in the absence of gpSid) would be to connect three gpN hexamers into a complete T=7 icosahedral face. The resulting shells will undergo maturation, involving expansion, processing of gpN, and removal of scaffold proteins. (From Marvik, O.J., Dokland, T., Nøkling, R.H., Jacobsen, E., Larsen, T., and Lindqvist, B.H., *J. Mol. Biol.,* 259, 59, 1995. By permission of the publisher Academic Press Ltd., London.)

5.III.B. SIPHOVIRIDAE

Long, noncontractile tails

Siphoviruses include at present five genera and about 2500 members. Heads are isometric or elongated. Tails measure 65–570 × 7–10 nm and are often flexible. The best-known member of this family is coliphage λ, but its properties and mode of replication are not representative of all viruses of the family. Lambda DNA has cohesive ("sticky") single-stranded ends which the virus uses to form a circular genome after infection. In the lytic cycle of the virus, synthesis of constitutive and lysis proteins is turned on and replication, after a stage of Θ rings, proceeds via a rolling-circle mode that generates multimeric DNA for encapsidation. In the lysogenic pathway, synthesis of lytic proteins is turned off and the circularized viral DNA is inserted into the host genome by a specific recombination event. The only other siphovirus whose developmental pathway has been illustrated in detail is *Vibrio* phage Φ149, a relative of coliphage T5.

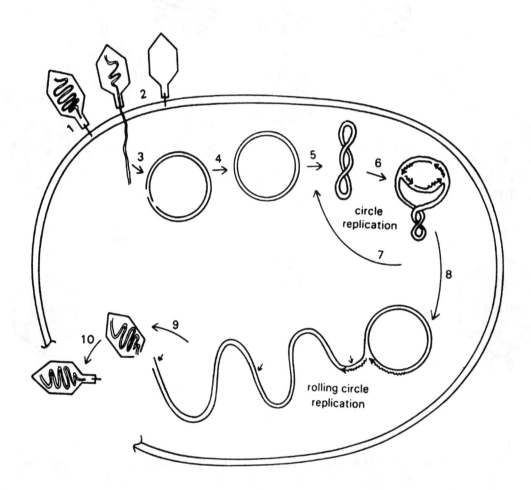

FIGURE 111

Replication of phage λ. (1) A phage attaches to a host cell and (2) injects its DNA which (3) circularizes by base pairing of complementary single-stranded ends. (4) The resulting nicked circle is closed by DNA ligase. (5) Supercoils are introduced by DNA gyrase. (6) DNA replication is first bidirectional and generates Θ structures which in turn (7) generate two daughter circles. (8) Subsequently, most phage DNAs replicate as rolling circles. (9) The concatemeric DNA thus produced is cut at *cos* sites (arrows) and packaged into phage heads. (10) Phage maturation is completed by addition of tails. (From Furth, M.F. and Wickner, S.H., in *Lambda II,* Hendrix, R.W., Roberts, J.W., Stahl, F.W., and Weisberg, R.A., Eds., Cold Spring Harbor Laboratory, Cold Spring Harbor, NY, 1983, 145. With permission.)

SIPHOVIRIDAE

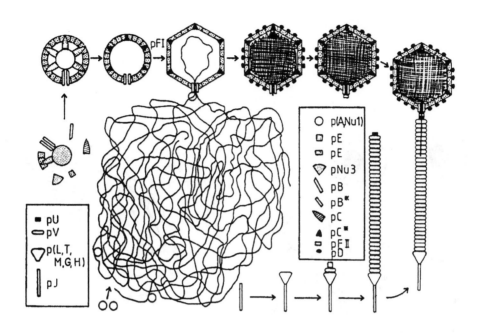

FIGURE 112

Morphogenesis of phage λ. A prohead is assembled, filled with DNA, and provided with a tail and tail fibers synthetized separately. The position of minor proteins (pB, pB*, pC, pNu3) is hypothetical. (From Hohn, T. and Katsura, I., *Curr. Topics Microbiol. Immunol.*, 78, 69, 1977. © Springer-Verlag, Berlin. With permission.)

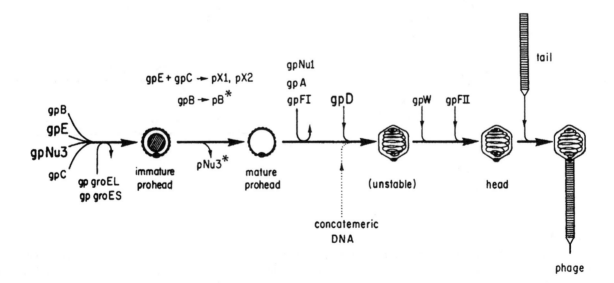

FIGURE 113

Assembly of phage λ. Phage-coded proteins must enter and travel through the pathway as indicated for proper assembly to occur. The tail is assembled independently and joins spontaneously to completed heads. Proteins gpD, gpE, and gpNu3 are denoted by large letters to indicate that large numbers of them are required for head assembly. (From Georgopoulos, C., Tilly, K., and Casjens, S., in *Lambda II*, Hendrix, R.W., Roberts, J.W., Stahl, F.W., and Weisberg, R.A., Eds., Cold Spring Harbor Laboratory, Cold Spring Harbor, NY, 1983, 279. With permission.)

SIPHOVIRIDAE

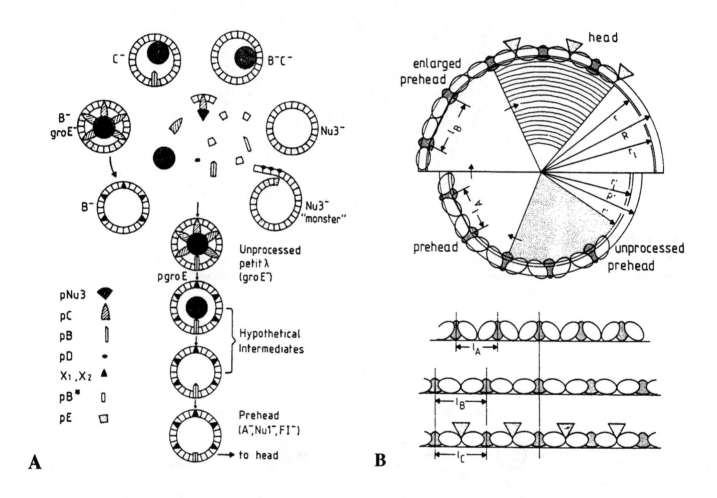

FIGURE 114

(A) Morphogenesis of λ proheads (*petite* particles). Locations of minor proteins pB, pC, pNu3, X1, and X2 are hypothetical. Most particles (upper flower) are considered to be abortive assembly products. Two types of particles (unprocessed prehead, prehead) are combined with two hypothetical intermediates in a pathway (stem). The existence of these intermediates has been inferred from the protein composition of abortive particles. (B) Hypothetical section through the λ head in various stages of assembly. Ellipsoids represent pE subunits, triangles represent pD trimers. In the lower part of the figure, lattice constants 1_A and 1_B (distance between centers of pE hexamers) are transposed as measured on polyheads analogous to icosahedral particles. r1 and r1*l*, lattice radii of preheads and heads as approximate spheres. (Adapted from Hohn, T. and Katsura, I., *Curr. Topics Microbiol. Immunol.*, 78, 69, 1977. © Springer-Verlag, Berlin. With permission.)

SIPHOVIRIDAE

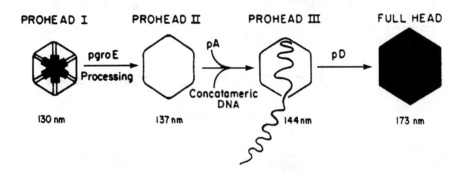

FIGURE 115

Assembly pathway of λ capsids. Prohead I, the earliest capsid precursor and only 130 nm in circumference, gives rise to a 5% larger, empty prohead II. Prohead III, an empty-appearing type of precursor about 144 nm in diameter, accumulates in certain lysates. The initial prohead expansion (prohead I to II) occurs concomitantly with the processing of capsid proteins, whereas latter expansions (prohead II to III and prohead III to full head) are probably associated with DNA packaging. (From Zachary, A. and Simon, L.D., *Virology,* 81, 107, 1977. With permission.)

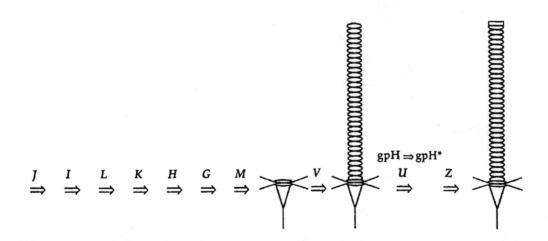

FIGURE 116

Assembly pathway of λ tails. Letters indicate tail genes and their sequence in tail assembly. In most cases, their action is probably equivalent to addition of the tail protein encoded by that gene to the nascent tail. The relative order of addition of the gpU protein and cleavage of gpH to gpH* has not been determined. (From Hendrix, R.W., *Curr. Topics Microbiol. Immunol.,* 136, 21, 1988. © Springer-Verlag, Berlin. With permission.)

SIPHOVIRIDAE

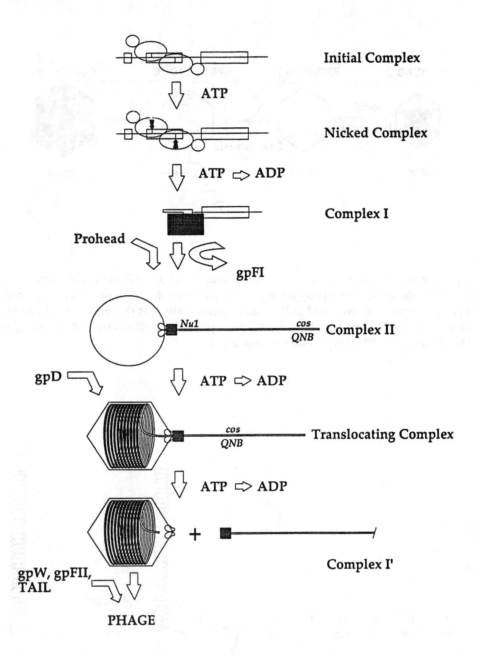

FIGURE 117

Model for λ DNA packaging. (1) An initial complex is formed comprising gpA (ellipse), gpNu1 (circle), *cosN* (small rectangle), and *cosB* (large rectangle). (2) Under stimulation by ATP, the initial complex becomes nicked in *cosN*. (3) Following ATP hydrolysis, complex I is formed by binding of terminase (grey box) to the left of the genome. (4) Prohead portal protein and gpA interact with the help of gpF1. (5) The prohead binds to complex I, generating complex II. (6) Packaging ensues under ATP hydrolysis. The prohead expands prior to packaging of 40% of DNA. (7) Packaging ends when the terminal *cos* is encountered and cleaved, a step which requires *cosQ*, a site at the left of *cosN*. The terminase of the packaging complex transfers to the left of the next genome to be packaged (complex I′). (8) The head is completed by addition of gpW, gpFII, and the phage tail. (From Catalano, C.E., Cue, D., and Feiss, M., *Mol. Microbiol.*, 16, 1075, 1995. With permission.)

SIPHOVIRIDAE

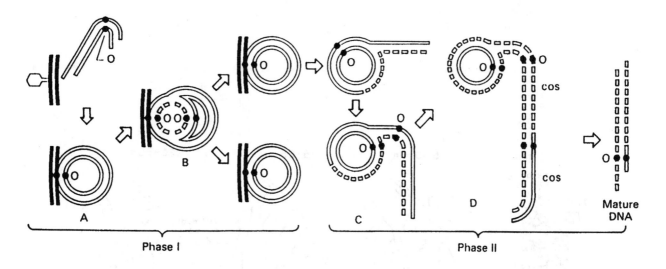

FIGURE 118

Replication of λ DNA. In phase I, the DNA injected by the infecting virion cyclizes (A) and, starting at a fixed origin (O), replicates symmetrically in association with the plasma membrane (B, double line). In phase II, the phage DNA, free of the plasma membrane, replicates asymmetrically by the rolling-circle model (C), generating linear concatemers. These are cut at *cos* sites (D) to generate mature molecules with cohesive ends. (From Dulbecco, R. and Ginsberg, H.S., in *Microbiology,* 3rd ed., Davis, B.D., Dulbecco, R., Eisen, H.N., and Ginsberg, H.S., Eds., Harper & Row, Hagerstown, MD, 1980, 919. With permission.)

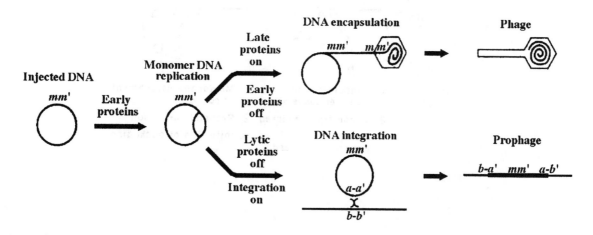

FIGURE 119

The two developmental pathways of phage λ. The injected DNA forms a covalently closed circle through pairing at the mature ends, m and m′, followed by ligation. After an early common phase, the phage may follow the productive or lysogenic pathways. (From Echols, H. and Murialdo, H., *Microbiol. Rev.,* 42, 577, 1978. With permission).

SIPHOVIRIDAE

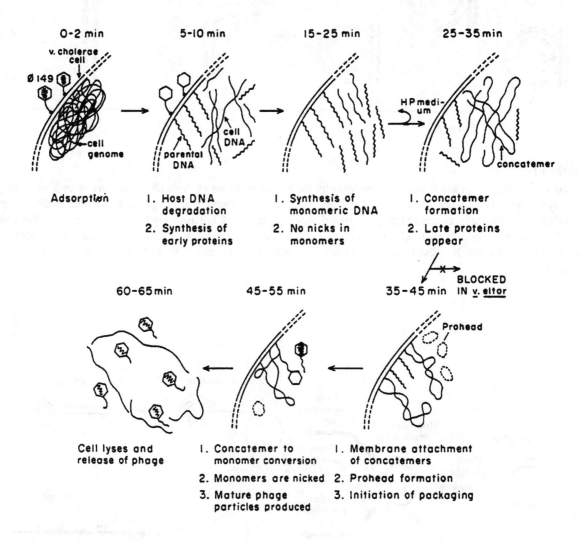

FIGURE 120

Replication of *Vibrio cholerae* phage Φ149. The cycle involves formation of concatemers and proheads. (From Majumdar, S., Dey, S.N., Chowdhury, R., Dutta, C., and Das, J., *Intervirology,* 29, 27, 1988. Reproduced with permission of S. Karger AG, Basel.)

5.III.C. PODOVIRIDAE

Short, noncontractile tails

Podoviruses presently comprise three genera and approximately 700 members. Tails measure about 20×8 nm. Well-known podoviruses include coliphage T7, which has DNA with staggered ends, and *Bacillus* phage ϕ29, whose DNA is circularized by a terminal protein as in adeno- and tectiviruses.

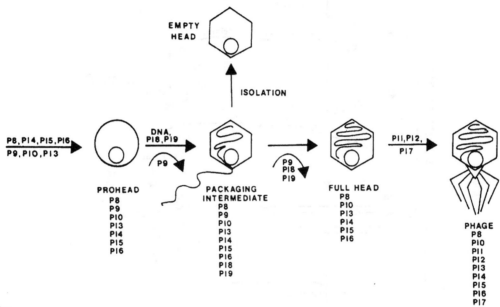

FIGURE 121

Coliphage T7 assembly. Five proteins come together to form a prohead. In the presence of P18, P19, and DNA, the prohead initiates DNA packaging. P9 exists as DNA enters. The head, which has started but not yet completed DNA packaging, is unstable and dissociates upon isolation into DNA and an empty head. After completion of packaging, the DNA is matured, P18 and P19 exit, and P11 and P12 attach to form a conical tail to which the tail fiber protein, P17, then attaches. (From Roeder, G.S. and Sadowski, P.D., *Virology,* 76, 263, 1977. With permission.)

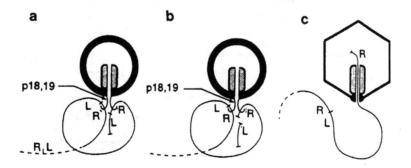

FIGURE 122

Initiation of T7 DNA packaging. (a) Capsid I, a spherical shell with an internal cylinder, binds to a DNA concatemer at two identical sites near the right end (R) of successive monomeric genomes within the concatemer. p18 and p19 are accessory proteins necessary for binding. (b) The mature right DNA end is formed by cleavage of the concatemer and release of the left end (L). (c) Capsid I converts to capsid II and the right DNA end enters the capsid. The fate of p18 and p19 is unknown. (From Son, M., Watson, R.H., and Serwer, P., *Virology,* 196, 282, 1993. With permission.)

PODOVIRIDAE

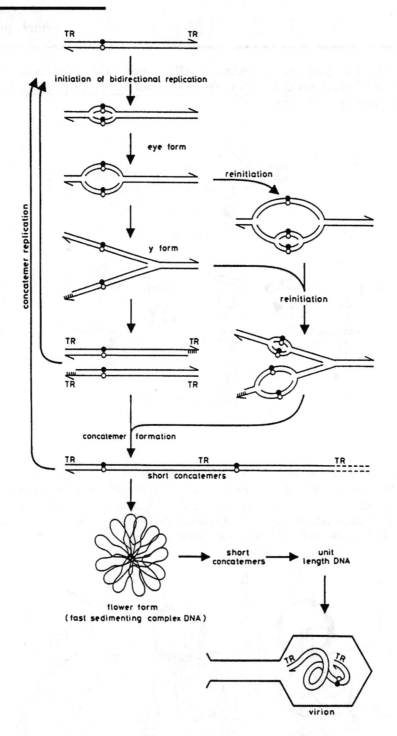

FIGURE 123

T7 DNA replication. The infecting DNA is linear and has terminal repeats (TR). Bidirectional replication of T7 DNA starts repeatedly at the primary origin of replication (black and white dots). Progeny molecules associate in head-to-tail fashion, probably through cohesive ends generated in the process, to give rise to short concatemers that are replicated again. A higher-order concatemeric structure (flower form) then appears. It is processed to short concatemers and finally to unit-length genomes. (From Keppel, F., Fayet, O., and Georgopoulos, C., in *The Bacteriophages,* Vol. 2, Calendar, R., Ed., Plenum Press, New York, 1988, 145. With permission.)

PODOVIRIDAE

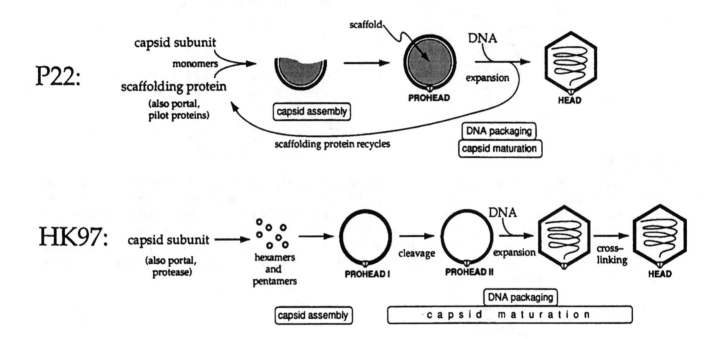

FIGURE 124

Capsid assembly pathways of *Salmonella* phage P22 and coliphage HK97. The two pathways are similar in overall design but differ substantially in detail (HK97 is a T1-like siphovirus; authors' note). (Reprinted from Hendrix, R.W. and Garcea, R.L., *Semin. Virol.*, 5, 15, 1994. By permission of the publisher Academic Press Ltd., London.)

PODOVIRIDAE

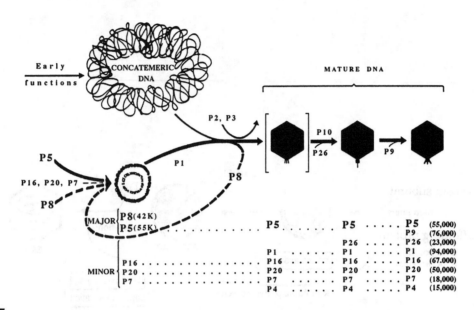

FIGURE 125

Pathway of P22 assembly and DNA encapsulation. A prohead with two major proteins (P5 and P8) and no DNA content is the first identifiable structure. It encapsidates and cuts a headful of DNA from the replicating concatemer with the help of P1, P2, and P3, while P8 exits from the particle without cleavage and catalyzes further rounds of prohead assembly. The filled head is first unstable (in brackets) and is stabilized by the actions of P10 and P26. The particle is completed by addition of a tail (P9). The proteins found in each structure are listed below it; their molecular weights are listed at the right. (Modified from Casjens, S. and King, J., *J. Supramol. Struct.*, 2, 202, 1974. ©1974 John Wiley & Sons. Reprinted by permission of Wiley-Liss, Inc., a subsidiary of John Wiley & Sons, Inc.)

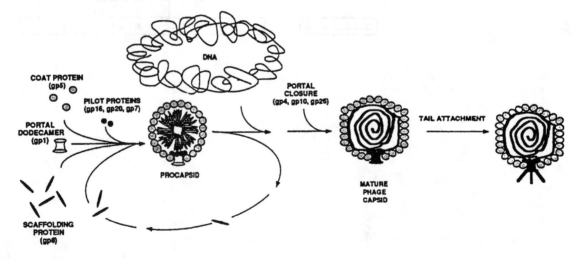

FIGURE 126

Pathway of P22 assembly. Coat protein (gp5) and scaffolding protein (gp8) copolymerize to form a procapsid. The portal vertex, a dodecamer of gp1, and the pilot proteins are incorporated at this point. The DNA enters through the portal vertex in a reaction requiring two packaging proteins (gp2, gp3). Simultaneously, the scaffolding protein exits and the coat protein lattice expands without addition of more proteins or modification of existing ones. The portal vertex is closed by three connector proteins (gp4, gp10, gp26) which stabilize the head, and a tail is added. (From Prasad, B.V.V. et al., *J. Mol. Biol.*, 231, 65, 1993. By permission of the publisher Academic Press Ltd., London.)

PODOVIRIDAE

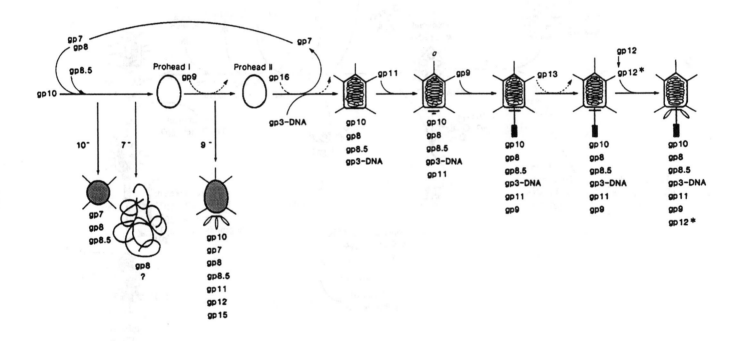

FIGURE 127

Morphogenetic pathway of φ29. Proteins composing a structure are listed below it. Prohead II and the structures below the main pathway (10⁻, 7⁻, 9⁻) are abortive side products. The composition of particle (a) has been inferred from morphology. (From Anderson, D. and Reilly, B., in *Bacillus subtilis,* Sonenshein, A.L., Hoch, J.A., and Losick, R., Eds., American Society for Microbiology, Washington, D.C., 1993, 859. With permission.)

PODOVIRIDAE

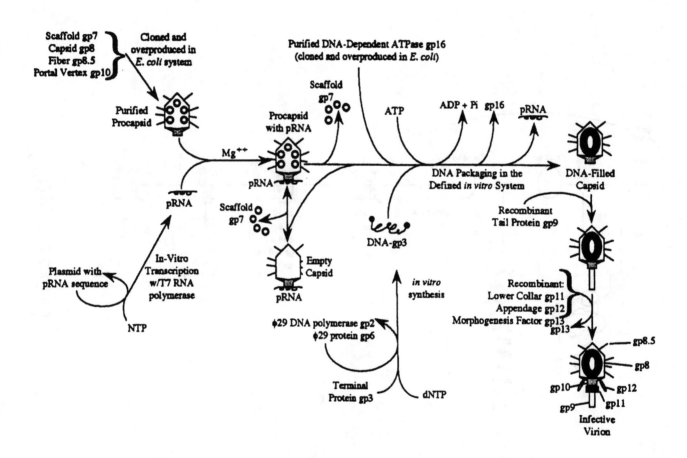

FIGURE 128

Assembly of infectious φ29 *in vitro*. The procapsid consists of the major capsid protein gp8 (about 200 copies), head fiber protein gp 8.5, scaffolding protein gp7 (180 copies), and portal vertex protein gp10 (12 or 13 copies). The scaffolding protein is necessary for procapsid assembly *in vivo*, but is not required for *in vitro* packaging of DNA. NTP, nucleoside triphosphate. (From Lee, C.-S. and Guo, P., *J. Virol.*, 69, 5018, 1995. With permission.)

PODOVIRIDAE

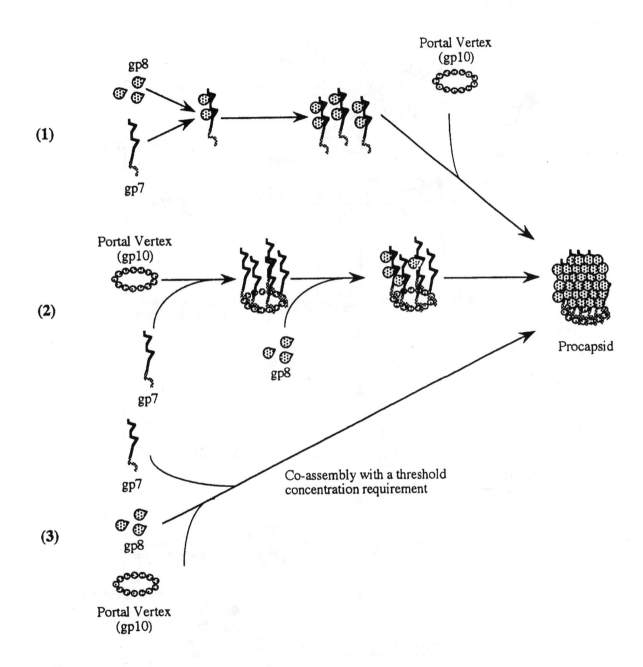

FIGURE 129

Three models for φ29 procapsid assembly. (1) Scaffolding protein gp7 and capsid protein gp8 interact first, forming heterodimers or hetero-oligomers. The gp7–gp8 complexes then interact with the portal vertex, consisting of 12 or 13 copies of gp10, to form active procapsids. (2) The portal vertex serves as initiator to interact with gp7, which bridges gp8 and gp10. The number of gp7 molecules recruited by gp10 could subsequently determine the number of gp8 molecules. Therefore, both gp10 and gp7 play roles in the determination of procapsid size and shape. (3) There are trimolecular interactions of gp7, gp8, and gp10 with a threshold concentration requirement. gp7 also serves as a bridge for gp10 and gp8; both gp7 and gp10 play roles in procapsid shape and size determination as in model 2. (From Lee, C.-S. and Guo, P., *J. Virol.*, 69, 5024, 1995. With permission.)

PODOVIRIDAE

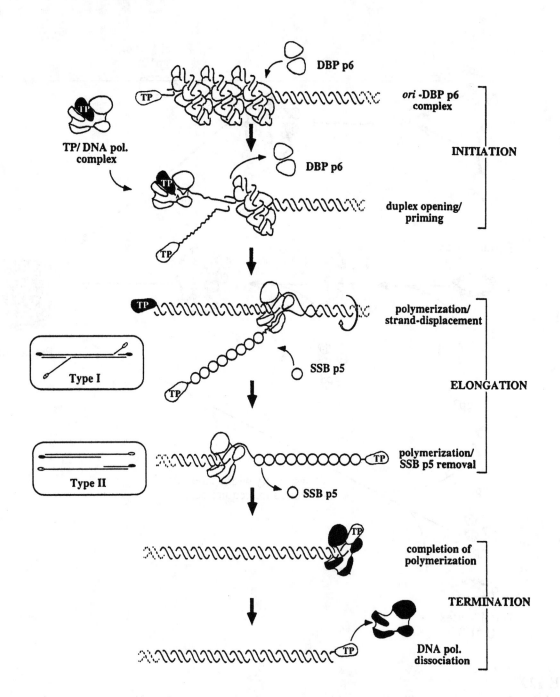

FIGURE 130

Protein-primed replication of φ29 DNA replication. Contrary to other tailed phages, φ29 DNA is covalently linked to a terminal protein which circularizes it and initiates replication (authors' note). φ29 DNA replication uses a mechanism of strand displacement and starts nonsimultaneously from either DNA end as in adenoviruses.[126] (Reprinted from Salas, M. et al., *FEMS Microbiol. Rev.*, 17, 73, © 1995. With kind permission of Elsevier Science–NL, Sara Burgerhartstraat 25, 1055 KV Amsterdam, The Netherlands.)

Chapter 6

VIRUSES WITH REVERSE TRANSCRIPTION

Four groups of viruses use reverse transcription to replicate their genomes: retro-, badna-, caulimo-, and hepadnaviruses. Retroviruses have RNA genomes able to integrate, after transcription into DNA, into the host chromosome. The other three groups contain DNA and are called pararetroviruses. The basic genetic structure of retro- and pararetroviruses includes a gene coding for structural proteins (*gag* type) followed by a gene coding for enzymatic functions (*pol* type). The templates for reverse transcription and the mRNA for viral proteins are RNA in retroviruses and pregenomic RNA in pararetroviruses, respectively.

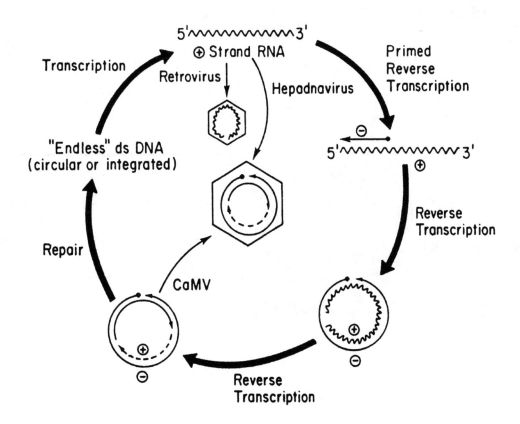

FIGURE 131

Nucleic acid packaging by retro- and pararetroviruses. The outside circle shows reactions undergone by unencapsidated nucleic acid; inside the circle are packaging reactions and reactions that occur within immature virions. RNA is represented by wavy lines and DNA by solid lines. Arrowheads indicate the direction of synthesis. The solid circle represents the primer for reverse transcriptase, a tRNA for retroviruses and a protein for the other viruses. (+) and (–) indicate message and anti-message sense strands, respectively. CaMV, cauliflower mosaic virus. (From Casjens, S., in *Virus Structure and Assembly,* Casjens, S., Ed., Jones and Bartlett Publishers, Boston, ©1985, 75. Reprinted with permission.)

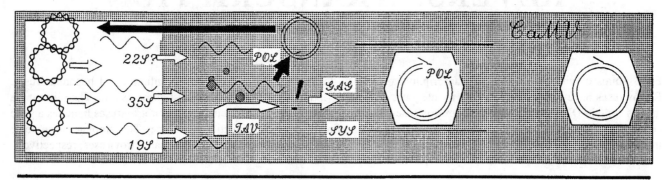

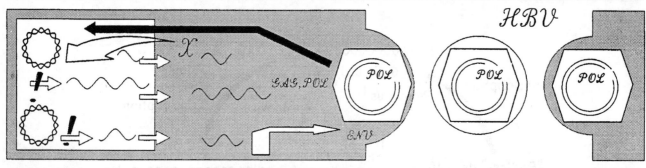

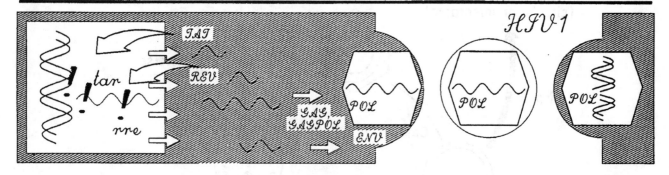

FIGURE 132

Control circuits of retro- and pararetroviruses. "!" symbolizes repression. Short open arrows indicate transcription and translation events, bent and kinked open arrows action of transactivators, and filled arrows transport of genomic RNA to the nucleus.
RNA and linear, open circular, and supercoiled forms of DNA are shown as self-evident symbols. Plant cells are stably arranged
and connected by plasmodesmata, which are modified by *SYS* protein to allow virus passage; animal (para)retroviruses are
enveloped while leaving cells and enter by fusion of the envelope with the cell membrane. *ENV, GAG, POL, REV, SYS, TAT,
TAV, X,* constitutive or regulatory viral proteins. CaMV, cauliflower mosaic virus; HBV, hepatitis B virus, HIV 1, human
immunodeficiency virus. (Reprinted from Hohn, T. and Fütterer, J., *Semin. Virol.,* 2, 55, 1991. By permission of the publisher
Academic Press Ltd., London.)

6.A. GENUS BADNAVIRUS

Circular dsDNA
Cubic, naked
Plants

Viruses of this "floating" genus are bacilliform particles derived from icosahedra. They usually measure 130 × 30 nm, but longer and shorter variants occur. Particles contain one molecule of circular DNA with site-specific single-stranded gaps. The sites of viral synthesis and assembly are unknown. The infecting viral genome is transcribed into RNA, which apparently serves as polycistronic mRNA for synthesis of viral proteins and for synthesis of DNA. Mediated by virus-specified reverse transcriptase, a (–) strand is made first and serves as a template for (+) strand progeny DNA. Novel viral strands are not ligated to form closed circles. Novel viruses accumulate in the cytoplasm.

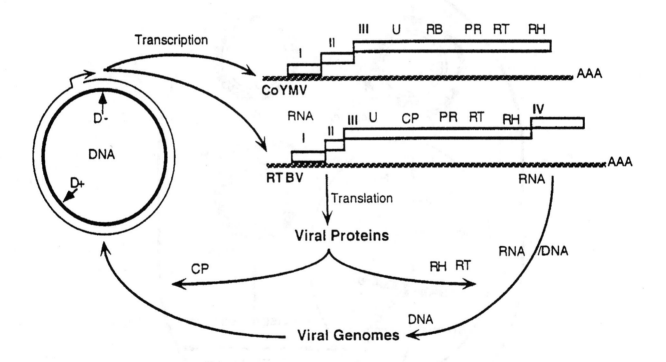

FIGURE 133

Genome organization and strategy of replication of *Commelina* yellow mosaic virus (CoYMV) and rice tungro bacilliform virus (RTBV). CP, coat protein; RB, aspartic protease; RB, RNA binding protein; RT, reverse transcriptase; RH, ribonuclease H; U, unknown function; D+, D, positive or negative DNA strands. (From Lockhart, B.E.L., Olszewski, N.E., and Hull, R., in *Virus Taxonomy, Sixth Report of the International Committee on Taxonomy of Viruses,* Murphy, F.A., Fauquet, C.M. et al., Eds., Springer, Vienna, *Arch. Virol.,* Suppl. 10, 185, 1995. © Springer-Verlag. With permission.)

6.B. GENUS CAULIMOVIRUS

Circular dsDNA
Cubic, naked
Plants

Viruses, typified by cauliflower mosaic virus (CaMV), infect dicotyledonous plants. Particles are about 50 nm in diameter and contain dsDNA in the form of an open circle with single-stranded gaps at fixed sites. Infecting DNA is converted in the nucleus to an episomal supercoiled "minichromosome," associated with host proteins, from which two mRNAs are transcribed. One species serves as a template for both translation into viral proteins and reverse transcription into progeny DNA. Viruses assemble in the cytoplasm, produce inclusion bodies, and are transmitted via plasmodesmata or by aphids.

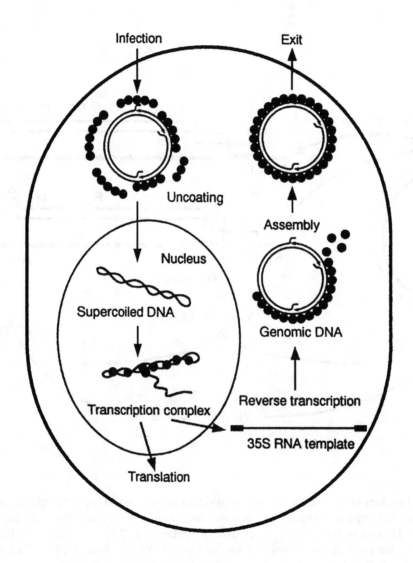

FIGURE 134
Replication cycle of CaMV, showing uncoating of the virus in the cytoplasm, formation of minichromosomes of viral DNA and host protein in the nucleus, and reverse transcription and virus assembly in the cytoplasm. (From Matthews, R.E.F., *Fundamentals of Plant Virology,* Academic Press, San Diego, 1992, 164. With permission.)

CAULIMOVIRUS

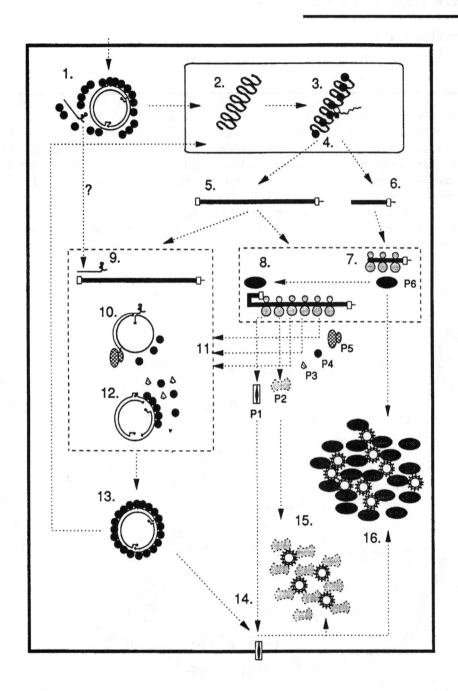

FIGURE 135

Detailed view of CaMV replication. (1) Uncoating of the virus liberates twisted DNA and a strong-stop primer. (2) DNA is supercoiled in the nucleus and (3) assembled into a transcriptionally active minichromosome. (4) Transcripts move into the cytoplasm. (5) A major transcript, 35S RNA, serves two roles. (6) 19S RNA is the messenger for P6. (7) Production of P6 precedes (8) translational transactivation of 35S RNA. (9) 35S RNA is selected for replication, possibly by strong-stop DNA. (9, 10) 35S is reverse transcribed to generate progeny virion DNA which is assembled into virions. (11, 12) Assembly requires P5 (reverse transcriptase) and structural proteins P3 and P4. (13) Mature progeny virions can move, mediated by P1, to adjacent cells through plasmodesmata (14) or are assembled into one or two types of inclusion bodies: those containing P2 (15) and destined for transmission by aphids and those containing P6 (16) which might function to regulate overflow of progeny virions. (Reprinted from Covey, S.N., *Semin. Virol.*, 2, 151, 1991. By permission of the publisher Academic Press Ltd., London.)

6.C. HEPADNAVIRIDAE

Circular supercoiled dsDNA
Cubic, enveloped
Vertebrates

This small family includes two genera of host-specific viruses of mammals and birds. Particles are about 45 nm in diameter and consist of an envelope, an icosahedral capsid, and one molecule of partially single-stranded DNA. Reverse transcriptase is associated with the virion. Human hepatitis B viruses are called "Dane particles" after their discoverer. The blood of infected people carries large quantities of particulate surplus envelope material, called "Australia" or hepatitis B surface antigen (HBsAg). Viruses enter hepatocytes by unknown mechanisms. Infecting DNA is converted in the nucleus to a covalently closed, circular, supercoiled, double-stranded molecule. The DNA is then made into viral proteins and RNA pregenomes or may become integrated into host cell DNA. RNA pregenomes are packaged into preformed capsids and reverse transcribed into (–) DNA, which serves as a template for (+) progeny DNA. Viruses acquire envelopes by budding into the endoplasmic reticulum and are excreted into the extracellular space.

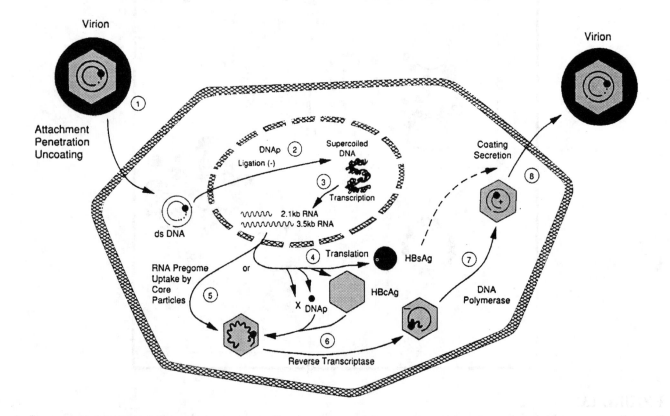

FIGURE 136

Replication of hepatitis B virus. The viral capsid enters a hepatocyte. The DNA is then uncoated and completed by DNA polymerase (DNAp) and a ligase to form supercoiled DNA. The latter is transcribed into two major RNA forms (2.1 kb and 3.5 kb), the larger of which (the RNA pregenome) can either be transcribed into viral proteins (HBsAG, HBcAg, HBxAg, DNAp) or packaged into capsids (core particles) and reverse transcribed into (–) strand DNA, which serves as a template for (+) strand DNA. Core particles are then provided with an envelope consisting of HBsAg and excreted. (From Hoofnagle, G.H. and Di Bisceglie, A.M., in *Antiviral Agents and Diseases of Man,* 3rd ed., Galasso, G.J., Whitley, R.J., and Merigan, T.C., Eds., Raven Press, New York, 1990, 415. With permission.)

HEPADNAVIRIDAE

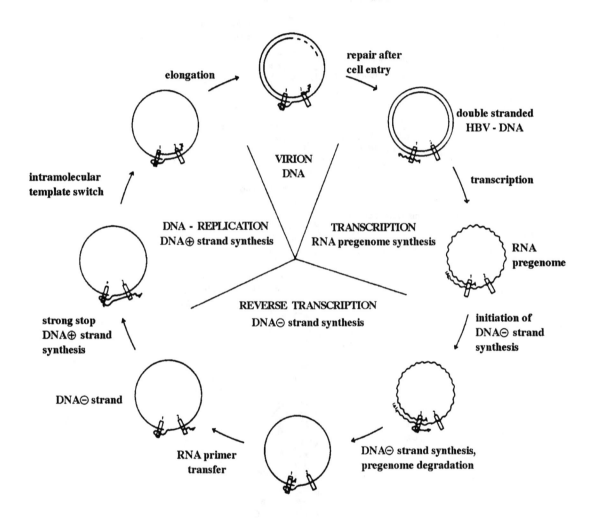

FIGURE 137

An alternative view of hepatitis B virus replication. (From Summers, J. and Mason, W.S., *Cell,* p. 403, 1982. © Cell Press. With permission.)

HEPADNAVIRIDAE

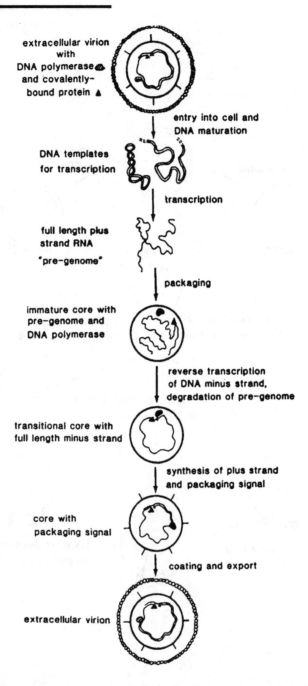

FIGURE 138

Replication of hepatitis B virus. The replication cycle is divided into three parts: transcription, reverse transcription, and DNA replication. After virus entry into the cell, the partially single-stranded viral DNA is completed to dsDNA which serves as a template for transcription. A terminally redundant RNA pregenome is then synthethized. DNA (–) strand synthesis starts at the 5′ end or close to the 3′ end of the RNA pregenome and a DNA (–) strand with a 5- to 8-nucleotide terminal redundancy is synthetized. The RNA pregenome is then degraded and a DNA (+) strand is produced. The terminal redundancy of the DNA (–) strand is probably used to facilitate an intermolecular template switch to allow the DNA (+) strand to become elongated. Before completion of the (+) strand, a virion with partially single-stranded DNA is exported to the blood. This route of reverse transcription and DNA replication is fundamentally different from that of retroviruses. (From Will, H., Reiser, W., Pfaff, E., Büscher, M., Sprengel, R., Cattaneo, R., and Schaller, J., *J. Virol.,* 61, 904, 1987. With permission.)

HEPADNAVIRIDAE

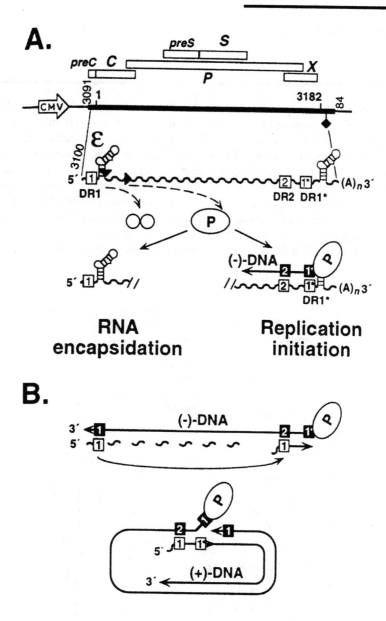

FIGURE 139

Basic aspects of hepatitis B virus (HBV) replication. (A) Genome organization and interactions of P protein with the pregenome. The solid black line represents a linear version of the HBV genome that in the virus is a partially circular molecule of 3182 nt (subtype ayW). Transcription of an authentic RNA pregenome can also be driven by foreign promoters (arrowhead), e.g., the cytomegalovirus (CMV) immediate–early promoter, from linear overlength DNA constructs. The diamond symbolizes the 3′-proximal polyadenylation signal. The open bars on the top show the four ORFs of HBV. The RNA pregenome (wavy line) starts at nt 3100 and, after one unit length, ends with a terminal redundancy of about 130 nt; hence, the direct repeat DR1 and ε (stem-loop structure) are present at both ends. The pregenome serves as mRNA for core (small spheres) and P protein whose binding to 5′ ε mediates RNA packaging. A 6-nt bulge in ε serves as a template for a short DNA primer that is transferred to DR1* to start replication. (B) Classical mode of HBV reverse transcription. Concomitantly with continuous (–) strand DNA synthesis, the RNA template is degraded except for the 5′-terminal oligonucleotide including 5′ DR1; this (+) strand DNA primer is transferred to DR2 and extended to the 5′ end of (–) strand DNA. A short terminal redundancy allows for a template switch to the 3′ end of (–) strand DNA and formation of the circular DNA genome. (From Nassal, M. and Rieger, A., *J. Virol.*, 70, 2764, 1996. With permission.)

6.D. RETROVIRIDAE

Dimeric linear (+) sense ssRNA
Cubic, enveloped
Vertebrates

This large family has seven genera. Its members infect mammals, birds, and reptiles. Viruses are frequently associated with tumors, leukemias (type B, C, or D retroviruses of mammals and birds), anemias, and immunodeficiency (genus *Lentivirus*). Particles are spherical and 80–100 nm in size. They consist of an envelope, an icosahedral or spherical capsid, and a central nucleoid or core containing reverse transcriptase and a dimer of two identical ssRNA molecules. Retroviruses are acquired by infection or are inherited as endogenous proviruses. Infecting virions adsorb to specific receptors and enter cells by membrane fusion with or without endocytosis. Viral RNA is reverse transcribed in the cytoplasm into complementary (–) DNA, on which (+) DNA strands are synthetized. The resulting linear dsDNA is integrated into cellular DNA to form a provirus. Integration is a prerequisite for replication. The integrated provirus persists in the cell or is transcribed into novel viral RNA and several mRNA species. Synthesis of viral proteins involves the formation of polyproteins which are cleaved during maturation. Capsids assemble at the plasma membrane or within the cytoplasm (type A) and generally acquire their envelope by budding.

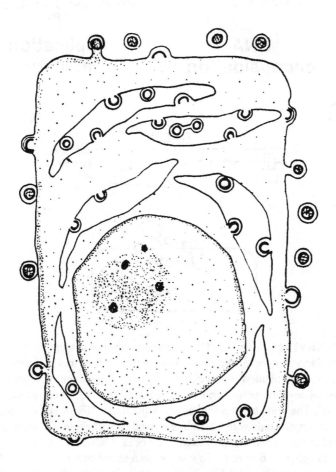

FIGURE 140
Cell showing replication of a mouse leukemia virus. Virus particles bud into cytoplasmic cisternae (the immature A type) or at the cell surface (A and C types). Nucleolar alterations are sometimes observed. (From Bernhard, W., in *Ciba Foundation Symposium on Cellular Injury,* De Reuck, A.V.S. and Knight, J., Eds., J. & A. Churchill, London, 1964, 209. With permission.)

RETROVIRIDAE

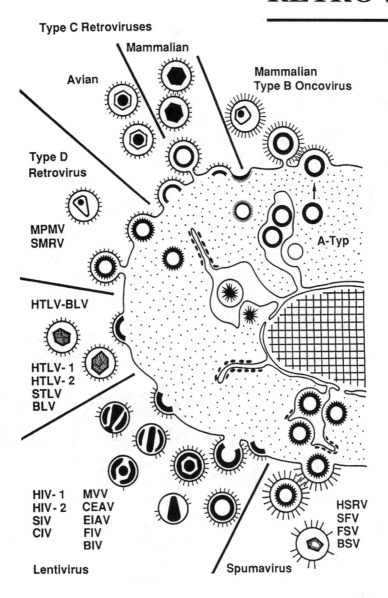

FIGURE 141

The retrovirus family. Types A–D correspond to the former *Oncornavirinae* subfamily. Type A particles are cores of B particles (exemplified by mouse mammary tumor virus). They migrate to the cell surface and become enveloped by budding, forming doughnut-shaped immature virions. Mature particles contain an eccentric, condensed, isometric core and are studded with surface projections (spikes) about 10 nm in length. Viruses of the C type cause sarcomas and lymphomas (chicken, cats, mice). Cores are assembled during the budding process. Mature C-type virions have 5-nm-long spikes and contain a central isometric core formed at the vicinity of the plasma membrane. D particles are found in certain primates and can also be related to malignancies. Preformed cores migrate to the plasma membrane and are enveloped by budding. Mature particles have elongated cores. HTLV-I (human T-cell leukemia virus) related agents represent an intermediate group with morphological features common to C-type and lentiviruses. They bud concomitantly with the assembly of the viral core and show surface knobs 58 nm in length. The electron-dense core of immature particles is transformed in mature viruses to a cone-shaped inner body. The pathogenic potential of spumaviruses is unknown. They occur frequently as "foamy virus" contaminants in cell cultures derived from organs such as kidneys. Spumaviruses show envelopment of preformed cores which, after release of the virions, never fully condense and are studded with surface projections. (From Gelderblom, H.R., Gentile, M., Scheidler, A., Özel, M., and Pauli, G., *AIDS Forsch.,* 8, 231, 1993; modified from Ref. 136. With permission of R.S. Schulz-Verlag, Starnberg-Percha, and Springer-Verlag, Vienna.)

RETROVIRIDAE

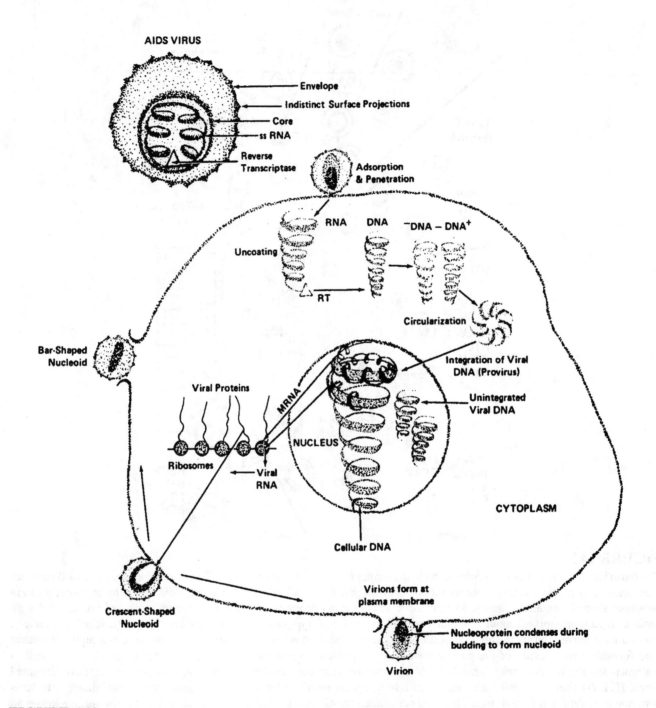

FIGURE 142

Overall events occurring during the replication of HIV. DNA complementary to viral RNA is made by reverse transcriptase. It then integrates as provirus to direct the synthesis of virus progeny. Particles are formed at the plasma membrane. There is no evidence of a nucleocapsid precursor. (From Palmer, E.L. and Martin, M.L., *Electron Microscopy in Viral Diagnosis*, CRC Press, Boca Raton, FL, 1988, 99. With permission.)

RETROVIRIDAE

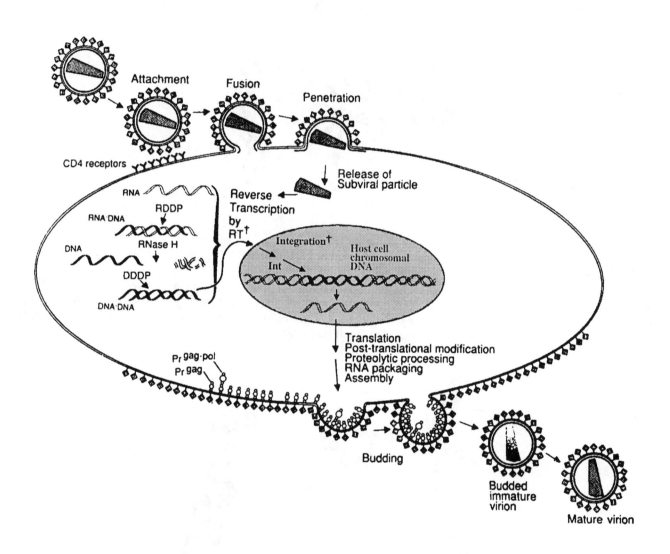

FIGURE 143

The HIV infectious cycle. The process is greatly oversimplified and the diagram is not meant to suggest that the various steps are discrete. The symbol +, for reverse transcription and integration, indicates that these processes are probably carried out in a capsid-like structure. (From Arnold, E. and Ferstandig Arnold, G., *Adv. Virus Res.,* 39, 1, 1991. With permission.)

RETROVIRIDAE

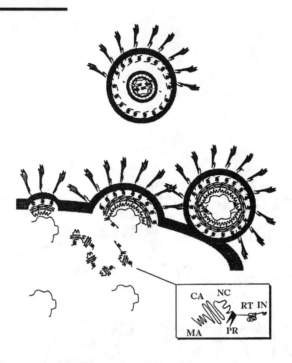

FIGURE 144

An assembly model applicable to most retroviruses (except types B and D). Cleavage of proteins and genome assembly do not take place until around the release of the virus. CA, capsid protein; IN, integrase; MA, gag protein; NC, RNA-associated "nucleocapsid" protein; PR, protease; RT, reverse transcriptase. (From Coffin, J.M., in *Virology,* 2nd ed., Vol. 2, Fields, B.N. and Knipe, D.M., Eds.-in-chief, Raven Press, New York, 1990, 1437. With permission.)

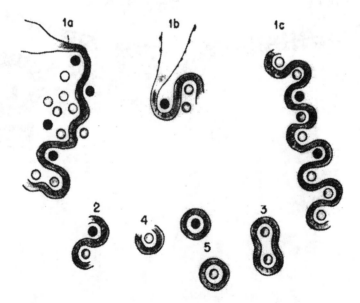

FIGURE 145

Origin of double envelopes in bovine syncytial virus (genus *Spumavirus*). Parts of the viral particles have double envelopes that are apparently acquired from the endoplasmic reticulum (ER). Nucleocapsids are observed free in the cytoplasm (1a), associated with the RER (1b), and budding in opposite directions across altered ER. Various intermediate stages (2, 3, 4) of double-enveloped virions (5) are illustrated. (From Boothe, A.D., Van der Maaten, M.J., and Malmquist, W.A., *Arch. Ges. Virusforsch.,* 31, 373, 1970. With permission.)

RETROVIRIDAE

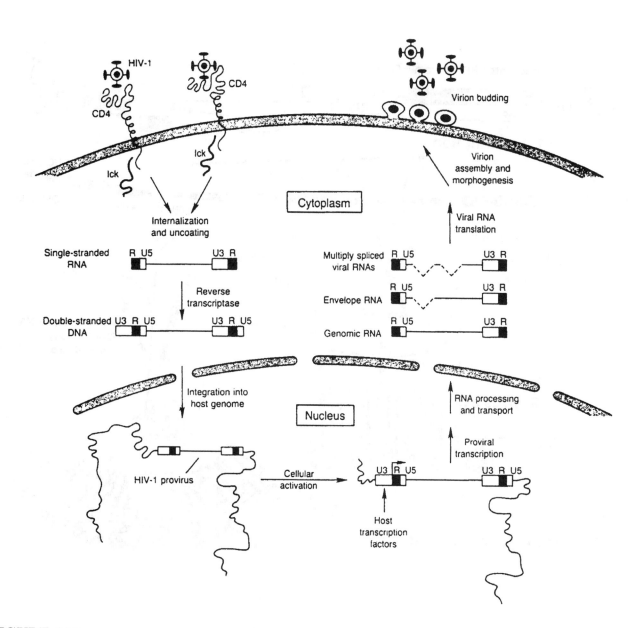

FIGURE 146

Life cycle of HIV-1. After interaction of gp120 with CD4 membrane receptors, gp41-mediated membrane fusion leads to the entry of HIV-1 (lck denotes a lymphoid-specific tyrosine kinase that binds to CD4). After uncoating, reverse transcription of viral RNA leads to production of dsDNA in the presence of appropriate host factors. HIV-1 integrase promotes the insertion of this viral duplex into the host genome, giving rise to the HIV-1 provirus. HIV-1 gene expression is stimulated initially by select inducible and constitutive host transcription factors with binding sites in the long terminal repeat of HIV-1, which leads to the sequential production of various viral mRNAs. The first RNAs produced correspond to the multiply-spliced species of about 2.0 kb encoding the Tat, Rev, and Nef regulatory proteins. Subsequently, viral structural proteins are produced, allowing the assembly of complete virions. Free HIV-1 virions are produced by budding and can reinitiate the retroviral life cycle by infecting other CD4+ target cells. (From Greene, W.C., *New England J. Med.*, 324, 308, 1991. © 1991 Massachusetts Medical Society. All rights reserved.)

RETROVIRIDAE

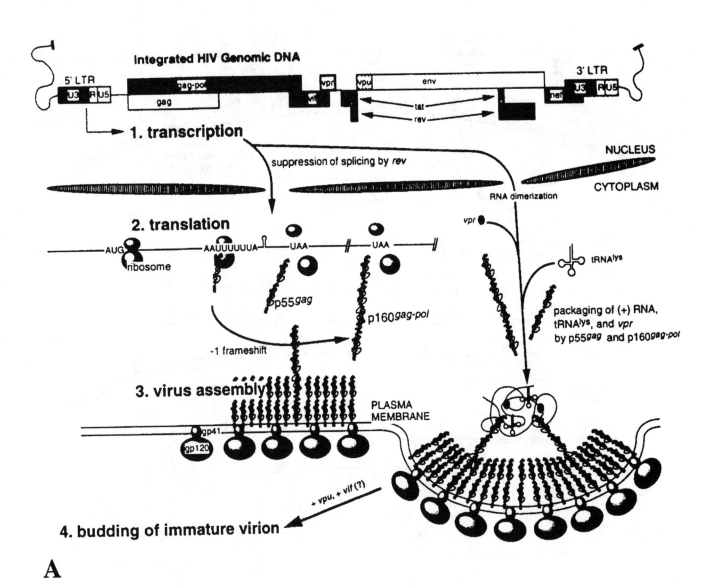

FIGURE 147

Detailed view of HIV-1 assembly and maturation. (1) Transcription starts at the repeat (R) region of the long terminal repeat (LTR). (2) Unspliced and singly spliced HIV RNA leave the nucleus for translation of structural proteins. (3) Envelope protein gp20 is transported to the plasma membrane prior to budding, followed by viral (+) RNA and other proteins. (4) Budding produces an immature virion. (5 and 6) The virion matures through proteolytic cleavage of polyproteins p55^{gag} and p160$^{gag-pol}$. (7) The mature capsid contains an assembly of RNA and proteins. Initiation of reverse transcription may actually start here. (Authors' legend.) (From Arts, E.J. and Wainberg, M.A., *Adv. Virus Res.,* 46, 97, 1996. With permission.)

RETROVIRIDAE

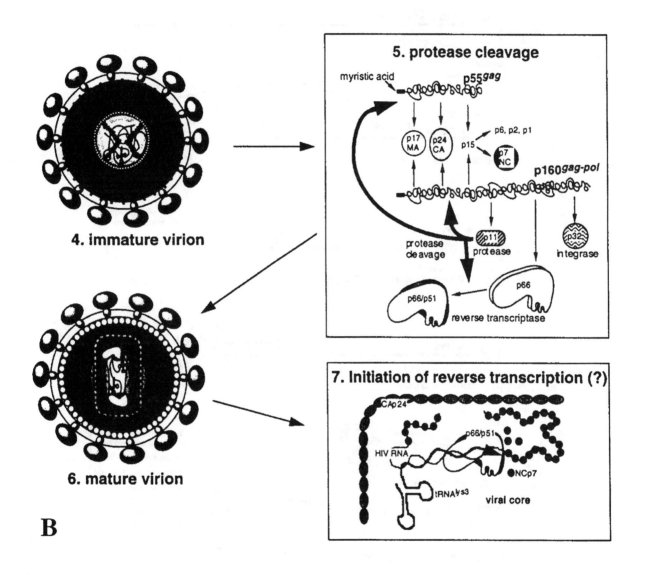

FIGURE 147 (continued)

RETROVIRIDAE

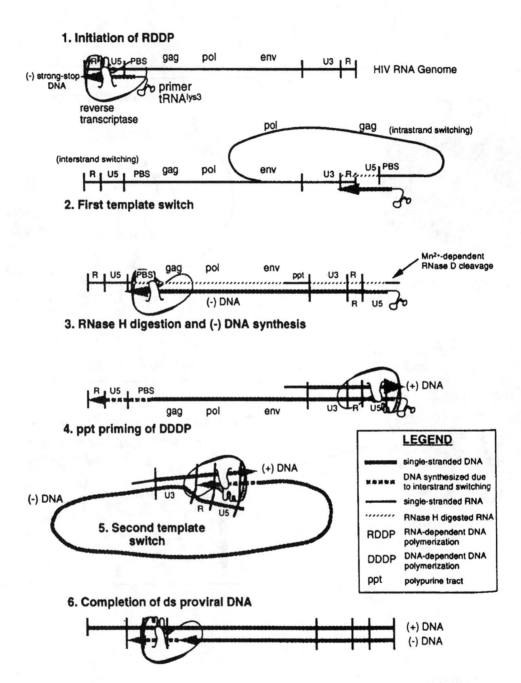

FIGURE 148

Reverse transcription of retroviruses. (1) RNA-dependent DNA polymerization (RDDP) starts. (2) (–) DNA is synthetized and RNA is digested by RNase H during (–) strand synthesis. (3) PBS (prime-binding sequence) is digested by RNase H during (–) strand synthesis. (4) A polypurine tract (ppt) is primed for completion of (+) DNA. (5) A second template switch primes DDDP for (6) completion of (+) DNA. (Authors' legend.) (From Arts, E.J. and Wainberg, M.A., *Adv. Virus Res.,* 46, 97, 1996. With permission.)

RETROVIRIDAE

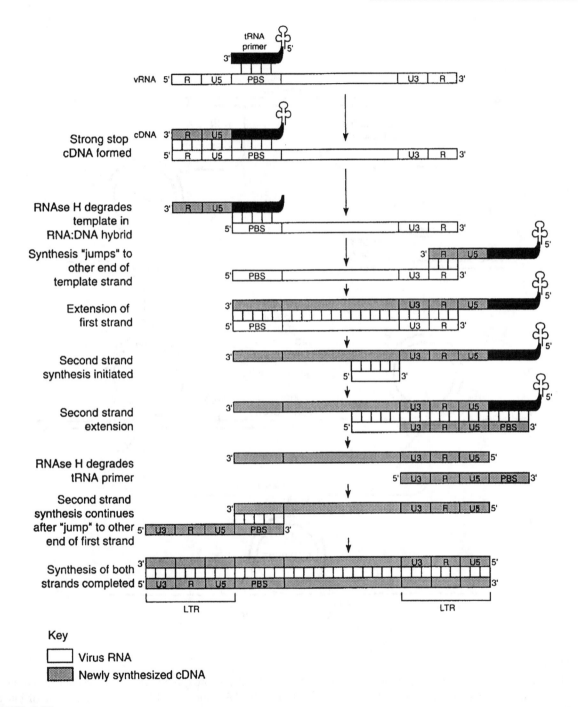

FIGURE 149

Mechanism of reverse transcription of retrovirus RNA genomes, in which two molecules of RNA are converted into a single, terminally redundant, double-stranded DNA provirus. (From Cann, A.J., Ed., *Principles of Molecular Virology,* Academic Press, London, 1993, 86. By permission of the publisher Academic Press Ltd., London.)

RETROVIRIDAE

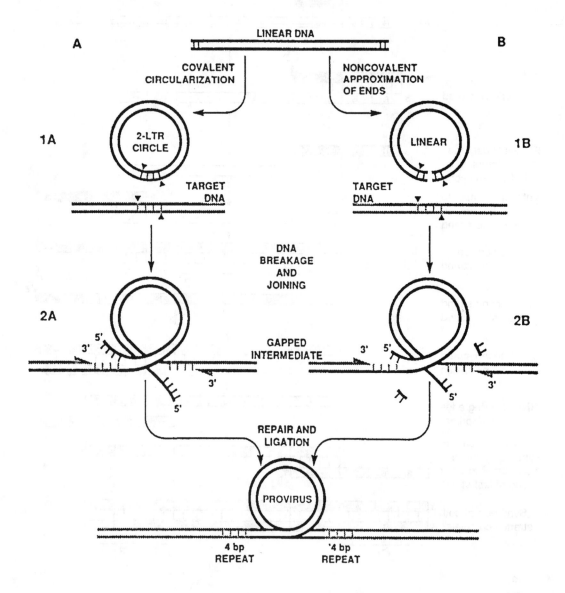

FIGURE 150

Two possible pathways (A and B) of integration of retroviral DNA. (1A) The proximal precursor to the integrated provirus is a 2-LTR circle, formed by ligation of the ends of a linear molecule. In (1B), linear viral DNA is the substrate of integration. In the following DNA breakage and joining steps, viral and target DNA are broken (arrowheads) and joined (3′ OH ends of viral DNA to 5′ P ends of target DNA). In the resulting intermediates (2A, 2B), the provirus is flanked by short gaps. Both pathways result in an integrated provirus with terminal repeats. The four base pairs of viral DNA that are ultimately lost upon integration (two from each end of the linear molecule) and the 4-bp target sequence that is duplicated upon integration are indicated by short lines perpendicular to DNA strands. (Authors' legend.) (From Brown, P.O., *Curr. Topics Microbiol. Immunol.,* 157, 19, 1990. © Springer-Verlag. With permission.)

RETROVIRIDAE

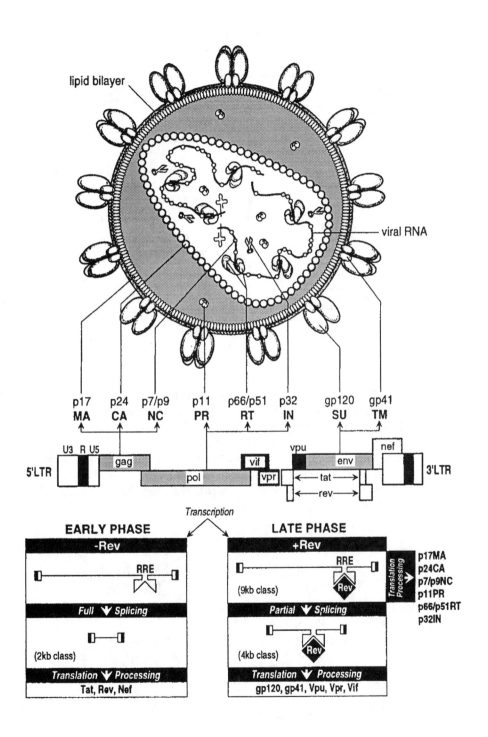

FIGURE 151

HIV-1 structure, genomic organization, and temporal gene expression. For abbreviations, see Fig. 144. (From Geleziunas, R., Bour, S., and Wainberg, M.A., *Adv. Virus Res.,* 44, 203, 1994. With permission.)

RETROVIRIDAE

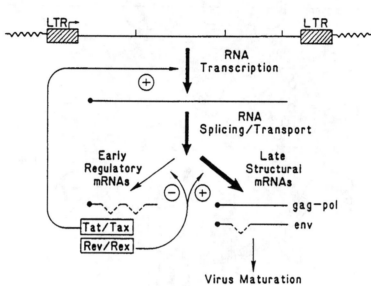

FIGURE 152

HIV-1 and HTLV-I gene regulation. "Early" viral mRNAs encode the viral regulatory proteins Tat and Rev (HIV-1) and Tax and Rex (HTLV-I). Tat and Tax stimulate the expression of all viral genes, at least at the level of transcription. This permits the expression of "late" structural gene for virus maturation and assembly. Transition from "early regulatory" to "late structural" is controlled by Rev and Rex, which regulate RNA transport and perhaps splicing. The sequence-specific action of Rev and Rex induces expression of the unspliced and singly-spliced viral mRNA species that encode the gag, pol, and env proteins, thus allowing for their translation. Simultaneously, Rev and Rex down-regulate the expression of multiply-spliced early regulatory mRNAs and thus inhibit their own production. (Reprinted from Greene, W.C. and Cullen, B.R., *Semin. Virol.*, 1, 195, 1990. By permission of the publisher Academic Press Ltd., London.)

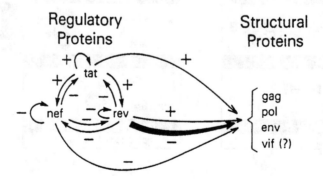

FIGURE 153

Regulatory circuit of HIV. *tat* and *nef*, respectively, regulate viral gene expression positively and negatively. *rev* activates expression of proteins at the expense of other regulatory proteins. When overexpressed (thick arrow), *rev* also down-regulates structural proteins. (From Wong-Staal, F., in *Virology*, 2nd ed., Vol. 2, Fields, B.N. and Knipe, D.M., Eds.-in-chief, Raven Press, New York, 1990, 1529. With permission.)

Chapter 7

RNA VIRUSES

I. (+) sense ssRNA Viruses

7.I.A. BROMOVIRIDAE

Linear (+) sense ssRNA, segmented
Cubic, naked, multipartite
Plants

This family consists of four genera. In three of them, virus particles are icosahedra 24–35 nm in diameter and constitute a tripartite system with three genomic and one subgenomic RNAs packaged into three different particles (RNA1, RNA2, RNA3 + subgenomic RNA). Members of the genus *Alfamovirus* are mostly bacilliform (30–57 × 18 nm) and constitute a quadripartite system with four RNA species. RNA1 and RNA2 encode single proteins each. RNA 3 has two cistrons encoding two proteins. Particles assemble in the cytoplasm and inclusion bodies are sometimes formed.

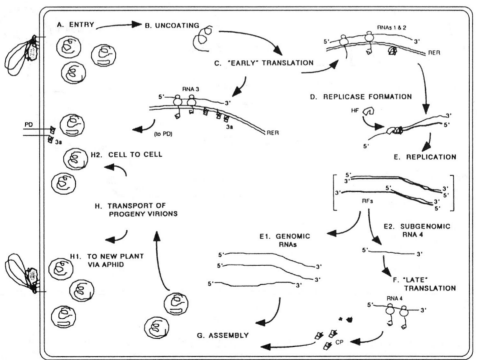

FIGURE 154

Life cycle of cucumber mosaic virus (CMV). (A) Viral particles enter the cell via an aphid vector. (B) Viral RNA is uncoated. (C) RNAs 1–3 are translated early in infection, possibly on membrane-bound ribosomes. (D) Translation products of RNAs 1 and 2 form the viral components of the replicase complex. (E) Replication occurs on membrane-associated replication complexes, which also contain host factor(s). Plus-sense RNAs are generated via a putative (ds?) replicative form. (E1) Genomic RNAs are generated for packaging into viral particles. (E2) Subgenomic RNA 4 is produced. (F) Translation of RNA 4 yields the coat protein. (G) Viral coat protein and genomic RNAs assemble to virions. (H) Progeny virus is transported to either another plant virus via an aphid vector (H1) or a new cell via plasmodesmata. It is unknown whether movement or protein-mediated transport of unencapsidated RNAs occurs. RER, rough endoplasmic reticulum; HF, host factor(s); RFs, replicative forms; PD, plasmodesmata. (From Palukaitis, P., Roossinck, M.J., Dietzgen, R.G., and Francki, R.I.B., *Adv. Virus Res.*, 41, 281, 1992. With permission.)

7.I.B. CORONAVIRIDAE

Linear (+) sense ssRNA, nonsegmented
Helical, enveloped
Vertebrates

Viruses infect mammals and are often associated with respiratory or enteric disease. Particles are roughly spherical, measure 160–180 nm in diameter, and contain a single RNA molecule. The genus *Coronavirus* is characterized by a "crown" of large club-shaped surface projections. Nucleocapsids seem to have icosahedral shells. Particles of the *Torovirus* genus are disk-, kidney-, or rod-shaped and have tubular nucleocapsids. Members enter cells through membrane fusion and replicate in the cytoplasm. Viral RNA is copied to form a template for synthesis of a nested set of subgenomic RNAs with common 3′ ends. Maturing viruses bud through the endoplasmic reticulum and Golgi membranes, but are not thought to mature at the plasma membrane.

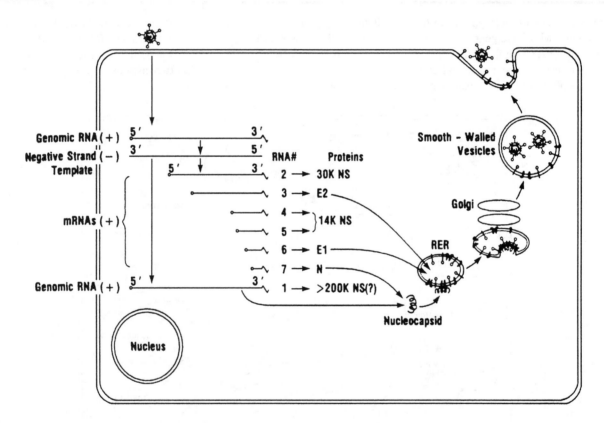

FIGURE 155

Coronavirus replication. Numbers of mRNAs and locations of nonstructural (NS) proteins may vary for different coronaviruses. Virions bind to the cell membrane and enter by membrane fusion or endocytosis. Viral genomic RNA acts as mRNA to direct the synthesis of viral RNA-dependent RNA polymerase. This enzyme copies the viral genomic RNA to form full-length (–) strand templates. These templates are copied to form new (+) strand genomic RNA, an overlapping series of subgenomic mRNAs, and leader RNA. All mRNAs are capped and polyadenylated and form a nested set with common 3′ ends. Each mRNA codes for a single polypeptide. The N protein binds to novel viral RNA to form helical nucleocapsids. E1, E2, and E3 glycoproteins are produced on membrane-bound polysomes. Some coronaviruses do not encode E3. Coronaviruses that encode E3 cause hemadsorption in infected cells. Virions are formed by budding at membranes of the Golgi apparatus and the RER, but not at the plasma membrane. Virions are released by cell lysis or by fusion of post-Golgi, virion-containing vesicles with the plasma membrane. (From Sturman, L.S. and Holmes, K.V., *Adv. Virus Res.,* 28, 35, 1983; updated version published in Ref. 148. With permission.)

7.I.C. FLAVIVIRIDAE

**Linear (+) sense ssRNA, nonsegmented
Cubic, naked
Vertebrates, invertebrates**

This family has three genera, *Flavivirus* (group B arboviruses including the agents of yellow fever and dengue), *Pestivirus* (pathogens of domestic animals), and human hepatitis C virus. Particles are spherical, 40–60 nm in diameter, and consist of an envelope surrounding an icosahedral capsid and a single RNA molecule. Flaviviruses were formerly considered as members of the *Togaviridae* family. Viruses enter cells by endocytosis. Multiplication occurs in the cytoplasm. The viral RNA replicates via a (–) strand intermediate and codes for a polyprotein which is cleaved into structural and nonstructural proteins. Flaviviruses do not produce subgenomic mRNA. Capsids assemble near cytoplasmic membranes and bud into cytoplasmic vesicles. Viruses exit via vacuoles or the secretory pathway.

a. PRIMARY TRANSLATION b. RNA REPLICATION

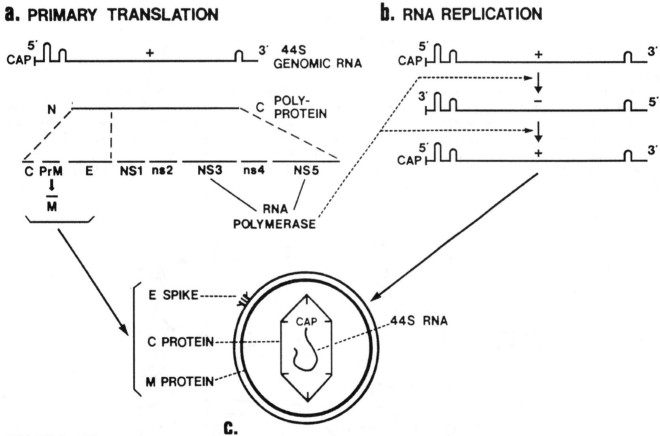

FIGURE 156

Flavivirus replication. (a) Genomic RNA is translated into a polyprotein. After or concomitant with proteolytic cleavage of the polyprotein, three polypeptides (PrM, E, NS1) associate with membranes, whereas capsid protein (C) accumulates in the cytoplasm. (b) Genomic RNA is replicated into complementary (–) strand RNA by RNA polymerase consisting of NS3 and NS5. Likewise, (–) strand RNA is replicated into progeny genomes. Secondary structures (Ω) at the 5′ and 3′ ends of genomic RNA and (–) stranded RNA may serve as recognition sites for RNA polymerase for replication and possibly for macromolecules involved in translation. (c) Progeny RNA is encapsidated within a capsid composed of C protein, which in turn buds to acquire an envelope. E protein trimers constitute spikes on the external surface of envelopes, and M protein, a nonglucosylated cleavage product of PrM, is located on the internal surface. (From Fenger, T.W., in *Textbook of Human Virology*, 2nd ed., Belshe, R.B., Ed., Mosby-Year Book, St. Louis, 1991, 74. With permission.)

FLAVIVIRIDAE

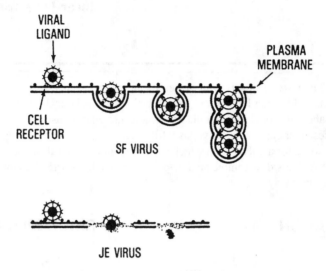

FIGURE 157

Comparative entry modes of togaviruses and flaviviruses. At physiological pH, the Semliki Forest (SF) togavirus enters by receptor-mediated endocytosis via plasma membrane invaginations and cytoplasmic vesicles. At pH 5.8, SF virus also enters by direct penetration of the cell surface. The Japanese encephalitis (JE) flavivirus penetrates the cell surface and disintegrates at or near the adsorption site. It is concluded that (a) at physiological pH, the fusion protein of SF virus is inactive and needs to be activated by acidic pH within the endosome, and (b) the fusion protein of JE virus is in an active state and capable of dissolving the host plasma membrane immediately after virus attachment. (From Hase, T., Summers, P.L., and Houston Cohen, W., *Arch. Virol.,* 108, 101, 1989. With permission).

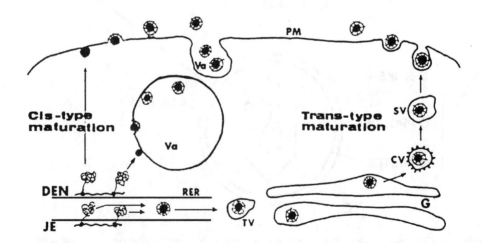

FIGURE 158

Cis-type and *trans*-type maturation in Dengue 2 (DEN) togavirus and Japanese encephalitis (JE) virus, respectively. Dengue viruses bud from the cell surface or are released into cytoplasmic vacuoles; JE viruses are released in a secretory-type pathway via the cisternae of the RER and various types of vesicles. DEN, synthesis of dengue virus structural proteins on ribosomes and their release into the cytosol; JE, synthesis of JE virus proteins and their release into cisternae. CV, coated vesicle; G, Golgi apparatus; PM, plasma membrane; RER, rough endoplasmic reticulum; SV, secretory vesicle; TV, transfer vesicle; VA, cytoplasmic vacuole. (From Hase, T., Summers, P.L., Eckels, K.H., and Baze, W.B., *Arch. Virol.,* 96, 135, 1987. With permission.)

7.I.D. LEVIVIRIDAE

Linear (+) sense ssRNA, nonsegmented
Cubic, naked
Bacteria

This family comprises two genera with about 90 members and is restricted to enterobacteria, acinetobacters, caulobacters, and pseudomonads. Particles are quasi-icosahedra about 24 nm in diameter and contain a single RNA molecule. Leviviruses have structural but apparently no phylogenetic relationships to picornaviruses and icosahedral ssRNA viruses of insects or plants. Viruses adsorb to sides of bacterial pili and probably reach the bacterial cell wall by pilus retraction. Replication proceeds via double-stranded intermediates. There is no polyprotein synthesis as in picornaviruses. Capsids seem to assemble around phage RNA. Novel viruses are released by cell lysis.

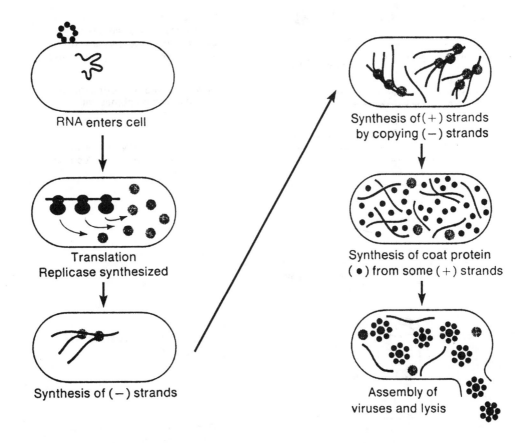

FIGURE 159

The levivirus life cycle. (Adapted from Freifelder, D., *Molecular Biology, A Comprehensive Introduction to Prokaryotes and Eukaryotes,* Science Books International, Boston, 1983, 659. © 1983 Boston: Jones and Bartlett Publishers. Reprinted with permission.)

LEVIVIRIDAE

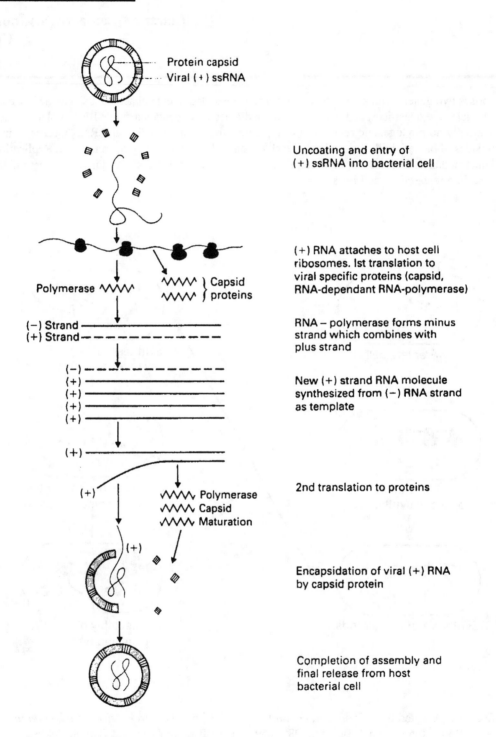

FIGURE 160

Detailed view of the levivirus life cycle (coliphage f2). (From Horne, R.W., *The Structure and Function of Viruses,* Edward Arnold, London, 1978, 31. Courtesy of R.W. Horne. With permission).

LEVIVIRIDAE

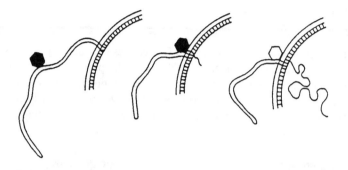

FIGURE 161

Pilus-dependent infection. The phage adsorbs to the side of a pilus and a short segment of phage RNA is extruded. This step is RNase-sensitive. Subsequent pilus retraction brings the phage in contact with the bacterial cell wall and phage RNA enters the cell. (From Beumer, J., Hannecart-Pokorni, E., and Godard, C., *Bull. Inst. Pasteur,* 82, 173, 1984. With permission).

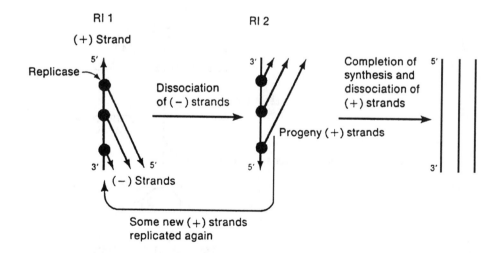

FIGURE 162

Replication of coliphage Qβ. RI, replication intermediates. (From Freifelder, D., *Molecular Biology, A Comprehensive Introduction to Prokaryotes and Eukaryotes,* Science Books International, Boston, 1983, 660. ©1983 Boston: Jones and Bartlett Publishers. Reprinted with permission).

7.I.E. NODAVIRIDAE

Linear (+) sense RNA, segmented
Cubic, naked
Vertebrates, invertebrates

This family represents a small group of icosahedral insect and fish viruses containing two (+) sense RNA molecules, thought to be encapsidated within a single capsid 30 nm in diameter. RNA1 encodes protein A and a putative RNA polymerase. RNA2 encodes protein α, the precursor of the capsid protein. Both are required for infection. Labile provirions have capsids composed of 180 α subunits. Following assembly of the viral shell, α subunits cleave spontaneously into mature coat proteins β and γ. Viruses replicate in the cytoplasm and frequently produce crystalline arrays. Capsids seem to assemble around the RNA.

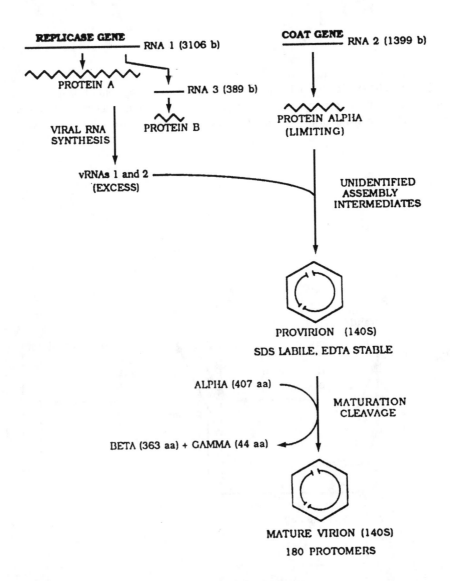

FIGURE 163

Nodavirus morphogenesis. Coat protein precursor α is synthetized in limiting amounts relative to virion RNAs 1 and 2. Maturation cleavage begins after provirion assembly and generates stable mature virions. Protomers refers to protein α or its cleavage product (β plus γ). (From Gallagher, T.M. and Rueckert, R.R., *J. Virol.,* 62, 3399, 1988. With permission.)

NODAVIDIDAE

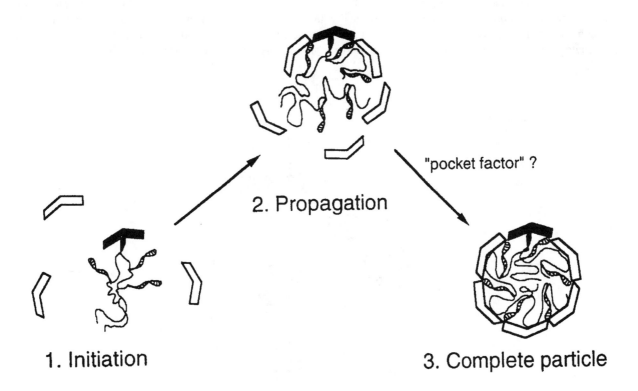

FIGURE 164

Involvement of RNA in nodavirus assembly. (1) The initiation step seems to be the formation of a nucleating complex in which a coat protein substructure, e.g., a monomer or multimer of asymmetric units ("platelet"), interacts with an encapsidation signal on the viral RNA. (2) "Propagation." Previous observations suggest that the initiation complex becomes a spherical particle by accretion of coat protein platelets which are guided into a growing shell by binding to helical duplex structures on the RNA. (From Schneemann, A., Gallagher, T.M., and Rueckert, R.R., *J. Virol.,* 68, 4547, 1994. With permission).

7.I.F. PICORNAVIRIDAE

Linear ssRNA, nonsegmented
Cubic, naked
Vertebrates

This family consists of five genera of mammalian viruses, including major human pathogens such as polioviruses (genus *Enterovirus*) and the agent of hepatitis A (genus *Hepatovirus*). Members of the *Cardiovirus* genus seem to infect both birds and insects. Particles are icosahedra 28–30 nm in diameter. Viruses are believed to enter cells through receptor-mediated endocytosis. Viral RNA replicates in complexes associated with cytoplasmic membranes. The viral genome is translated into a polyprotein precursor which is cleaved into structural and nonstructural proteins. Capsids seem to assemble around progeny RNA. Intracytoplasmic inclusion bodies are frequent. Viruses are released through vacuoles or by cell lysis.

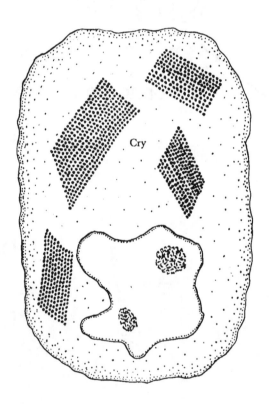

FIGURE 165

Cell showing poliovirus replication. Virions form crystal-like aggregates (Cry) in the cytoplasm. (From Bernhard, W., in *Ciba Foundation Symposium on Cellular Injury,* De Reuck, A.V.S. and Knight, J., Eds., J. & A. Churchill, London, 1964, 209. With permission.)

PICORNAVIRIDAE

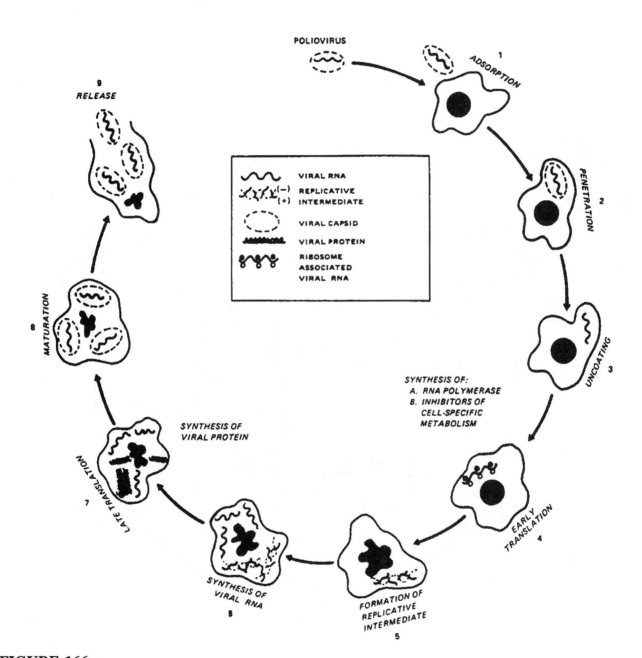

FIGURE 166

Replication of poliovirus. (1) Viruses adsorb to specific cellular receptors, (2) are taken in by viropexis, and (3) the viral RNA is uncoated. (4) The single-stranded genomic RNA serves as its own messenger RNA and is translated. (5) A replicative intermediate, a partially double-stranded molecule consisting of a complete RNA strand and numerous partially completed strands, is formed. (6) A replicative form (RF) of RNA is produced, consisting of a (+) and a (−) strand. The (+) strand may serve as a template for continued replication, may serve as mRNA for synthesis of structural proteins (7), or may become encapsidated. (8) Mature virus particles are released by cell lysis. (From Melnick, J.L., in *Virology and Rickettsiology,* Vol. I, Part 1, Hsiung, G.-D. and Green, R.H., Eds., CRC Press, Boca Raton, FL, 1978, 111. With permission.)

PICORNAVIRIDAE

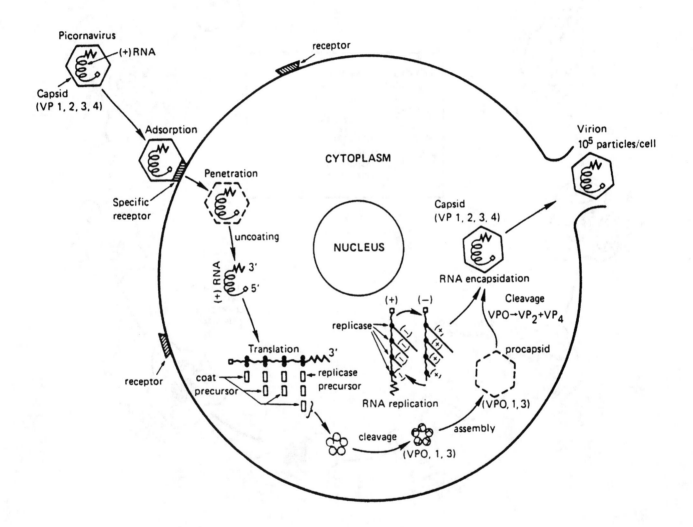

FIGURE 167

Picornavirus replication. Uncoated RNA acts as messenger to direct the synthesis of viral progeny. Viruses are usually released by lysis of cells. (From Palmer, E.L. and Martin, M.L., *Electron Microscopy in Viral Diagnosis,* CRC Press, Boca Raton, FL, 1988, 65. With permission.)

PICORNAVIRIDAE

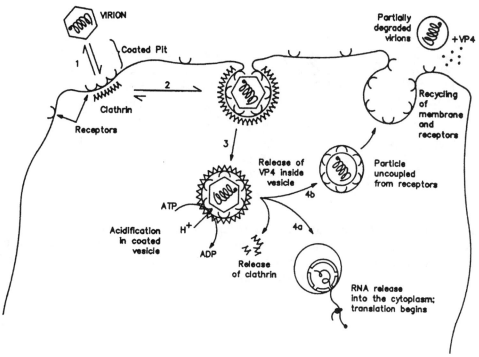

FIGURE 168

Receptor-mediated endocytosis in picornaviruses. Viruses adsorb to glycoprotein receptors. Clustering of receptors at clathrin-coated pits is followed by invagination and intake (endocytosis) to form clathrin-coated vesicles. Acidification inside these vesicles triggers release of VP4 and clathrin. Fusion of the lipid bilayer with hydrophobic patches within the acid-unfolded capsid protein presumably triggers release of viral RNA into the cytosol. Vesicle membranes and receptors are recycled. (From Rueckert, R.R., in *Virology,* 2nd ed., Vol. 1, Fields, B.N. and Knipe, D.M., Eds.-in-chief, Raven Press, New York, 1990, 507. With permission.)

FIGURE 169

"Docking" of a picornavirus to a hypothetical, pore-like membrane receptor with fivefold symmetry. The receptor is drawn to fit into the canyon surrounding the fivefold axis at a vertex of a virion. (From Crowell, R.L., *ASM News,* 53, 422, 1987. With permission.)

PICORNAVIRIDAE

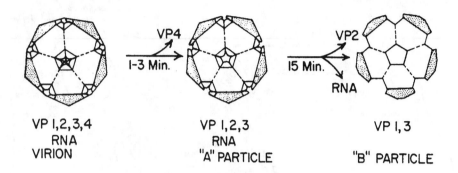

FIGURE 170

Schema for sequential release of viral polypeptides during uncoating. A VP4 pentamer is depicted in the center of each VP2 pentamer, considered to be located at each of the 12 vertices of the virion. It is suggested that interaction of the virion with specific receptors dissociates VP4 and VP2. (From Crowell, R.L. and Siak, J.-S., in *Perspectives in Virology,* Vol. 19, Pollard, M., Ed., Raven Press, New York, 1978, 39. With permission.)

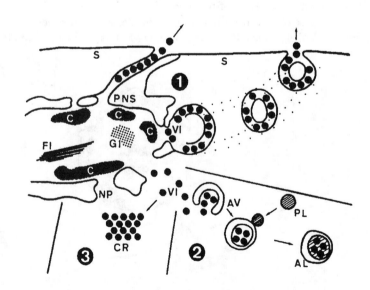

FIGURE 171

Terminal events in coxsackievirus infection. (1) Virus release via straight channels and complicated membrane formations. (2) Cellular defense reaction. (3) Formation of virus crystals. AL, autolysosomes; AV, autophage vacuole; C, chromatin; GI, granular inclusion; NP, nuclear protein; PL, primary lysosome; PNS, perinuclear space; VI, virions. (From Bienz, K., Bienz-Isler, G., Egger, D., Weiss, M., and Loeffler, H., *Arch. Ges. Virusforsch.,* 31, 251, 1970. With permission.)

PICORNAVIRIDAE

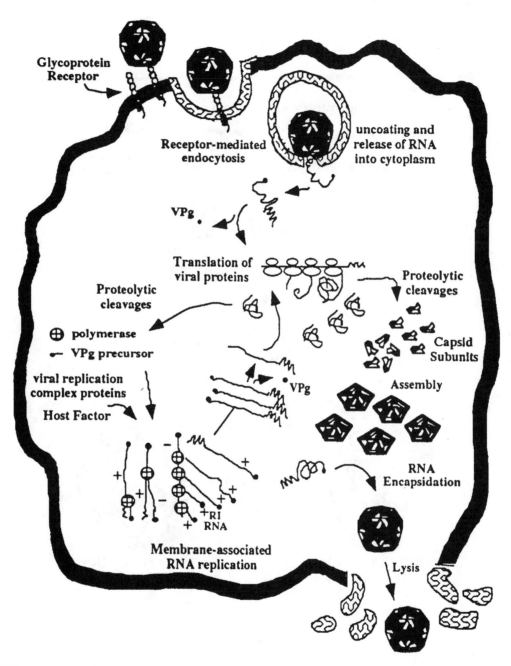

FIGURE 172

Poliovirus replication cycle. The virus binds to glycoprotein receptors and is believed to enter the cell by receptor-mediated endocytosis. The capsid undergoes conformational changes and viral RNA is released into the cytoplasm. The viral genome is translated on host ribosomes to generate proteins required for RNA replication and encapsidation of progeny RNA. Replication of RNA occurs in virus-induced, membrane-associated complexes of viral and host protein(s). Synthesis of (–) sense strands is followed by excess synthesis of (+) strand RNA. Structures in which several nascent (+) strands are simultaneously synthetized on the same (–) strand template are known as RI or replicative intermediates. Viral RNA is encapsidated and mature viral particles are released by cell lysis. (From Ansardi, D.C., Porter, D.C., Anderson, M.J., and Morrow, C., *Adv. Virus Res.*, 46, 1, 1996. With permission.)

PICORNAVIRIDAE

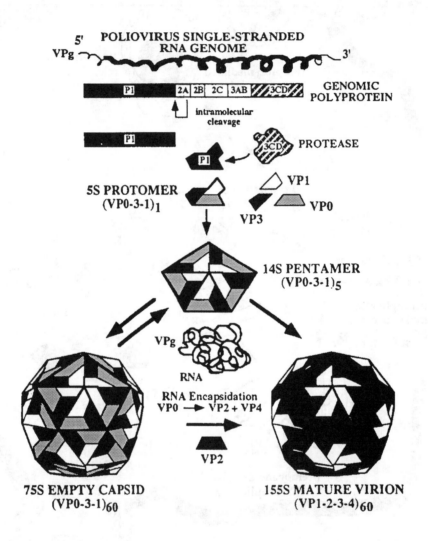

FIGURE 173

Pathway of poliovirus assembly. Capsid proteins are translated as part of a polyprotein and are released by cleavage by viral protease 2A as a 97-kDa precursor designated P1. The viral polyprotein 3CD cleaves P1 into capsid proteins VP0, VP3, and VP1. These proteins are believed to form a 5S protomer subunit. Five protomers assemble to 14S pentamers. Twelve pentamers form a 75S empty capsid (procapsid) by condensation around a nucleating RNA genome. On RNA encapsidation, VP0 is cleaved into VP2 and the inner capsid protein VP4. (From Ansardi, D.C., Porter, D.C., Anderson, M.J., and Morrow, C., *Adv. Virus Res.,* 46, 1, 1996. With permission.)

PICORNAVIRIDAE

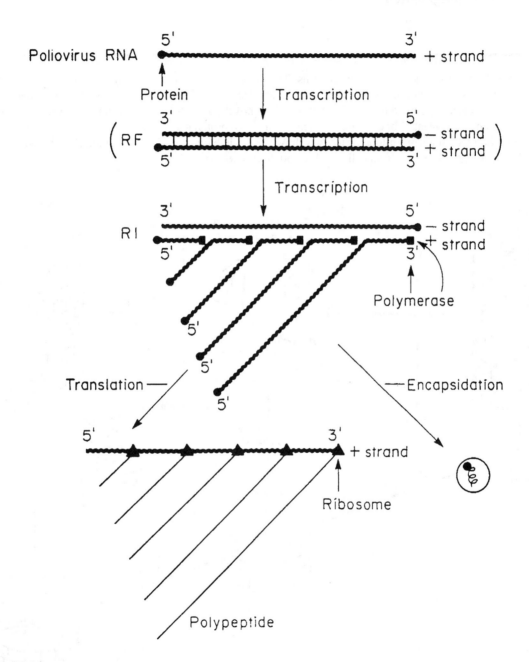

FIGURE 174

Replication and functioning of poliovirus RNA. Parental virus RNA (+) is transcribed into a polyprotein and, by a polymerase derived from the polyprotein, into (–) strands, to yield double-stranded replicative forms (RF) which probably exist only briefly. Progeny (+) strands are then described repeatedly from these (–) strands by a peeling-off mechanism. The structure consisting of a (–) stranded template and several (+) stranded transcripts in various stages of completion is known as the replicative intermediate (RI). Progeny (+) strands are either translated or encapsidated. (From Joklik, W.K., in *Zinsser Microbiology,* 17th ed., Joklik, W.K., Willett, H.P., and Amos, D.B., Eds., Appleton-Century-Crofts, New York, 1980, 1040. © Appleton & Lange. With permission.)

7.I.G. GENUS TOBAMOVIRUS

Linear (+) RNA, nonsegmented
Helical, naked
Plants

This "floating genus" comprises the tobacco mosaic virus (TMV), one of the first viruses whose structure and assembly were known in detail. Particles are rigid rods of 300 × 10 nm, constituted of two intertwined helices of coat protein and a single molecule of RNA. Viral RNA is copied into (–) RNA from which novel viral RNA and subgenomic RNAs are produced. The latter are translated into "movement" (for cell-to-cell-spread) and capsid proteins. Particle assembly starts with a nucleation disk of protein subunits into which a hairpin loop of RNA is drawn. Particle elongation is bidirectional, the 5′ end of RNA being coated first. Viruses are able to move from cell to cell through plasmodesmata.

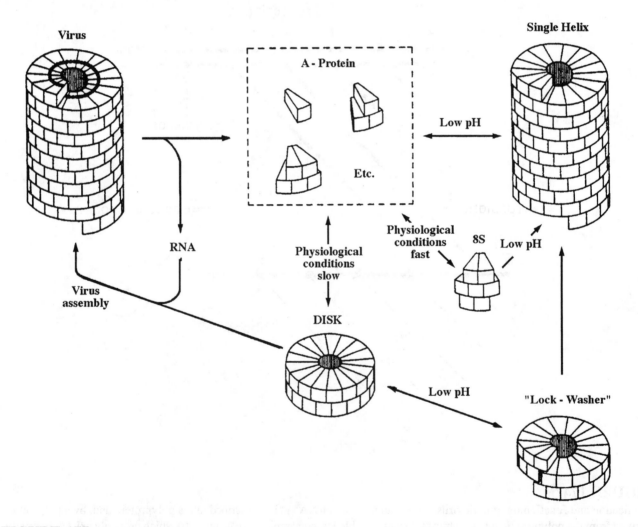

FIGURE 175

Structures and interconversions observed in some relatively well-characterized aggregates of TMV protein. (From Butler, P.J.G. et al., in *Structure–Function Relationship of Proteins, Proc. Third John Innes Symposium*, Markham, R. and Horne, R.W., Eds., North-Holland, Amsterdam, 1976, 101. With permission).

TOBAMOVIRUS

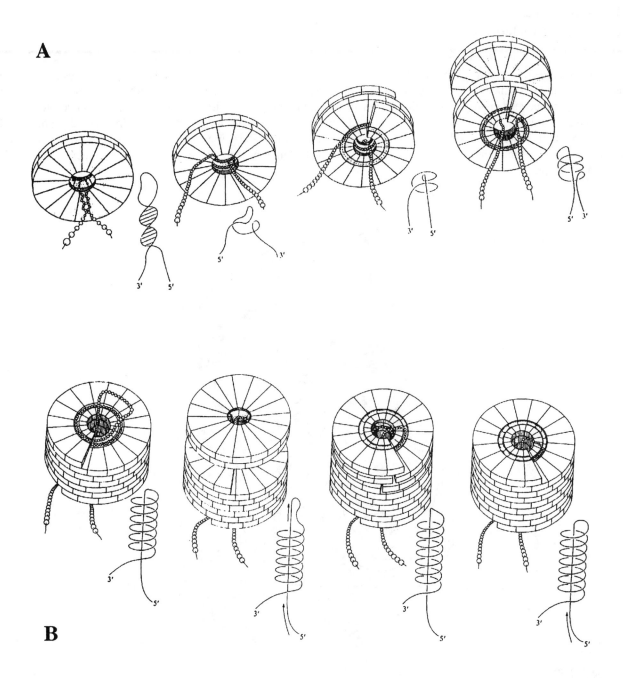

FIGURE 176

(A) Possible mechanism for nucleation of TMV assembly. The expected configuration of the RNA backbone at each stage is represented in the small diagrams. (B) Possible mechanism of particle elongation along the 5' RNA tail. The expected configuration of the RNA and its movement through the central hole of the rod are shown in the small diagrams. (From Butler, P.J.G., *J. Gen. Virol.*, 65, 253, 1984. With permission.)

7.I.H. TOGAVIRIDAE

Linear ssRNA, nonsegmented
Cubic, enveloped
Vertebrates, invertebrates

Togaviruses include the genera *Alphavirus* (type A arboviruses) and *Rubivirus* (human rubella virus). Particles are spherical, measure 60–70 nm in diameter, and consist of an envelope, an icosahedral capsid, and a single RNA molecule. Viruses enter cells by receptor-mediated endocytosis. Multiplication occurs in the cytoplasm. Viral (+) RNA produces a polyprotein precursor for nonstructural proteins and a (–) strand copy which acts as a template for novel viral DNA and subgenomic RNA encoding structural proteins. Capsids assemble from capsid protein and RNA in the cytosol. Viral glycoprotein spikes are inserted into cellular membranes. Capsids acquire envelopes by budding through intracellular and plasma membranes marked by glycoprotein spikes.

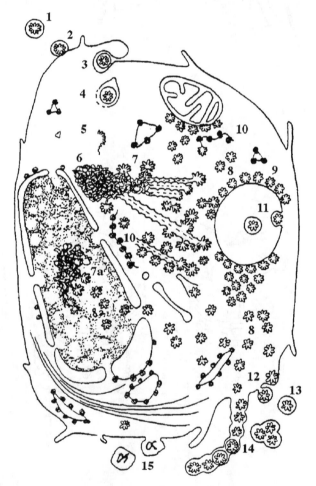

FIGURE 177

Morphogenesis of Venezuelan equine encephalitis virus. (1) An extracellular virion (2) attaches to a cell and (3, 4) enters it by viropexis. (5) Viral ribonucleoprotein is released and induces viral synthesis. (6, 7) Virus-specific structures, designated as "virus factories," appear in the cytoplasm and (7a) probably in the nucleus. It is assumed that they are the sites for (8) synthesis of viral RNA and protein and formation of viral nucleoids, which are localized (9) near polyribosomes or (10) near mitochondria. (11) Viruses are formed with participation of vacuole membranes or (12) the cell membrane. (13) Complete virions and (14, 15) anomalous forms are released from the cell. (From Bykovsky, H.F., Yershov, F.I., and Zhdanov, V.M., *J. Virol.,* 4, 496, 1969. With permission.)

TOGAVIRIDAE

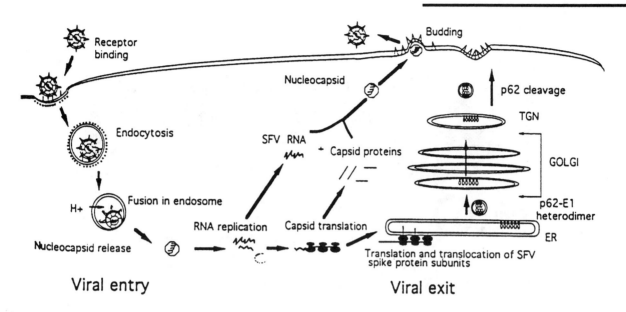

FIGURE 178

Life cycle of Semliki Forest virus, a prototypical alphavirus. ER, endoplasmic reticulum; TGN, *trans*-Golgi network. (From Kielian, M., *Adv. Virus Res.,* 45, 113, 1995. With permission.)

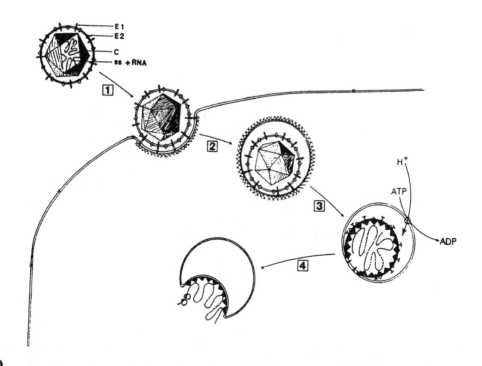

FIGURE 179

Entry of rubella virus. (1) After attachment to a receptor, (2) the virus is taken up and included in a coated vesicle. (3) The low pH in the lysosome induces the E1 protein to become fusogenic and facilitates fusion of virus and lysosome membranes. In parallel, the low pH induces a solubility shift of the capsid, allowing it to fuse with the lysosome membrane and causing intralysosomal uncoating. (4) Viral RNA is released upon fusion. (From Mauracher, C.A., Gillam, S., Shukin, R., and Tingle, A.J., *Virology,* 181, 773, 1991. With permission.)

TOGAVIRIDAE

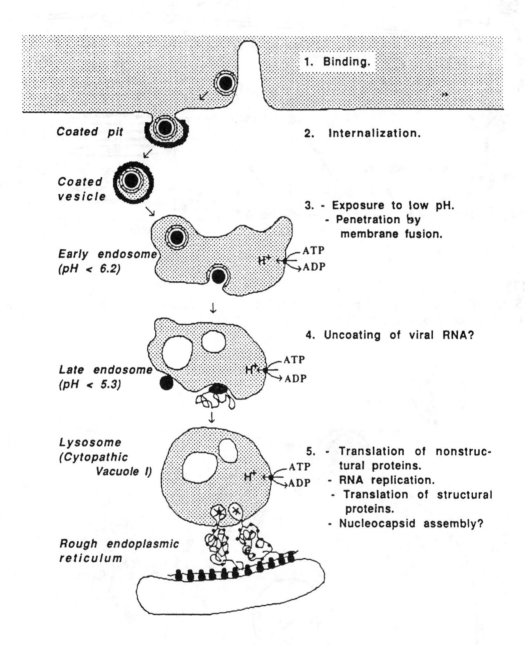

FIGURE 180

Entry of Semliki Forest virus. The virus attaches preferentially to microvilli, moves laterally to coated pits, and enters by receptor-mediated endocytosis. Final penetration by membrane fusion is triggered in the early endosome. Almost immediately after intake, the virus capsid is exposed to pH >6.2 and penetrates the cytoplasm. In virus mutants with a lower pH threshold of fusion or in cell mutants with defective endosomal acidification, virus capsids are released from late endosomes. Replication of viral RNA seems to occur at the cytoplasmic side of lysosomal membranes. Synthesis of viral RNA and of nonstructural and structural proteins, as well as nucleocapsid assembly, may occur in the same extensive area located between cytopathic vacuoles and lysosomes. (From Marsh, M. and Helenius, A., *Adv. Virus Res.*, 36, 107, 1989. With permission.)

TOGAVIRIDAE

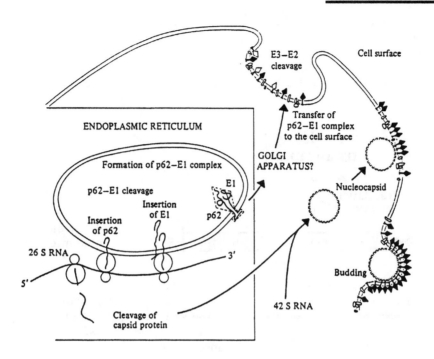

FIGURE 181

Assembly of Semliki Forest virus. Envelope proteins are inserted into the endoplasmic reticulum where protein p62 (precursor of E2 and E3) and the E1 protein form a two-chain structure. This is transported to the cell surface where the p62 protein is cleaved to E2 and E3. Protein C combines with viral 42S RNA in the cytoplasm to form the nucleocapsid, which diffuses to the cell surface to bind to a cluster of spike proteins to initiate the budding process. More spike proteins move into the budding patch and are immobilized by binding to the nucleocapsid. (From Simons, K. and Garoff, H., *J. Gen. Virol.,* 50, 1, 1980. With permission.)

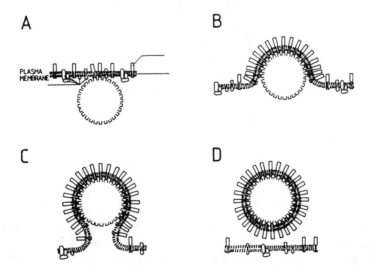

FIGURE 182

Possible mechanism of budding in alphaviruses. (A) The nucleocapsid recognizes and binds to the cytoplasmic domain of spike glycoprotein and the plasma membrane. (B, C) More spike glycoproteins are bound and a "bud" is formed. Host proteins are excluded from the spike glycoprotein patch. (D) A new virus particle is formed when all spike-binding sites on the nucleocapsid are filled. (Adapted from Simons, K., Garoff, H., and Helenius, A., in *Membrane Proteins and Their Interactions with Lipids,* Vol. 1, Capaldi, R., Ed., Marcel Dekker, New York, 1977, 207. With permission.)

TOGAVIRIDAE

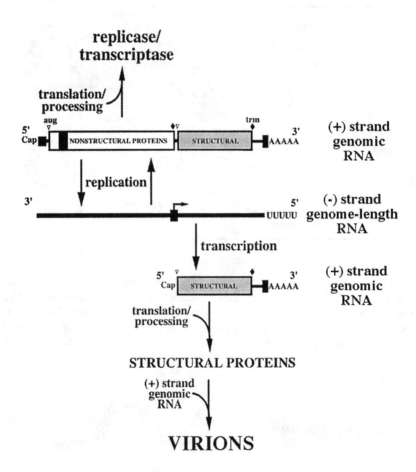

FIGURE 183

Alphavirus genome structure and replication. Translated regions of genomic and subgenomic RNAs are boxed and nonstructural and structural proteins are indicated. *Cis*-acting protein sequences important for replication and transcription are symbolized by small, shaded boxes, as is the sequence in the nonstructural region important for encapsidation (filled box). An arrow on the (–) strand RNA template signals the start site for subgenomic mRNA transcription. Empty triangles indicate translation initiation (aug) signals and black diamonds indicate termination signals (trm). (Redrawn from Bredenbeek, P.J. and Rice, C.M., *Semin. Virol.,* 3, 297, 1992. By permission of the publisher Academic Press Ltd., London.)

7.I.I. GENUS TYMOVIRUS

**Linear (+) sense RNA, nonsegmented
Cubic, naked
Plants**

Viruses of this group infect dicotyledonous plants. Particles are icosahedra 30 nm in diameter and comprise two types, B and T particles. Only B particles, which contain molecules of RNA, are infectious. Tymovirus replication is essentially the same as in tobamo- and togaviruses; for example, coat protein is transcribed from subgenomic mRNA. Viral RNA seems to replicate within vesicles at the periphery of chloroplasts and to assemble with coat protein subunits inserted into the chloroplast membrane.

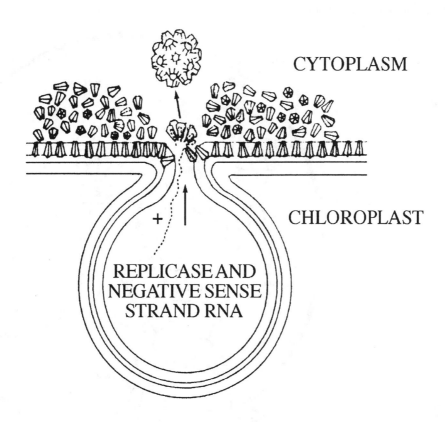

CYTOPLASM

CHLOROPLAST

+

REPLICASE AND NEGATIVE SENSE STRAND RNA

FIGURE 184

Assembly of turnip yellow mosaic virus. Pentamer and hexamer clusters of coat protein subunits are synthetized by the endoplasmic reticulum and accumulate in the cytoplasm overlaying the chloroplast. Subunit clusters are then inserted into the outer chloroplast membrane in an oriented fashion, that is, with the hydrophobic sides that are normally buried in the complete protein shell lying within the lipid bilayer of the chloroplast membrane. An RNA strand synthetized or being synthetized within a vesicle emerges through the vesicle neck. A specific sequence in the RNA recognizes or binds a feature of pentamer clusters lying in the outer chloroplast membrane or near the vesicle neck. This initiates virus assembly, which proceeds by addition of pentamers and hexamers. Completed virus particles are released into the cytoplasm. (From Matthews, R.E.F., *Plant Virology,* 3rd ed., Academic Press, San Diego, 1991, 238. With permission.)

II. (−) sense ssRNA Viruses

7.II.A. ARENAVIRIDAE

Linear (−) sense ssRNA, segmented
Helical, enveloped
Vertebrates

This family with a single genus of essentially rodent-pathogenic viruses includes the agent of Lassa fever. Particles are enveloped, roughly spherical, and usually 110–130 nm in diameter. The envelope contains two molecules of RNA circularized by addition of protein, RNA polymerase, and a variable number or ribosomes. Viruses attach to receptors and enter cells via endosomes. Replication and assembly take place within the cytoplasm without obvious cytopathic effects. RNA molecules are ambisense. The process of replication/translation involves subgenomic mRNA species, but is not fully understood. In the late stages of infection, virus-specified glycoproteins are inserted into the plasma membrane. Viruses mature by budding at insertion sites and include cellular ribosomes in the process.

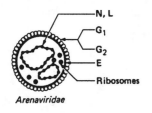

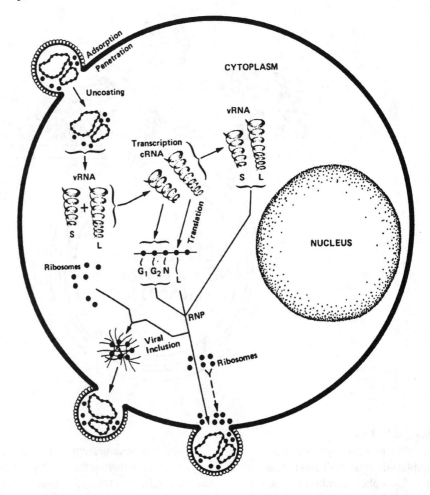

FIGURE 185

Replication of arenaviruses. The virus envelope fuses with the plasma membrane and its contents, viral nucleoprotein and ribosomes, are released into the cell. Transcription, translation, and replication of viral RNA take place in the cytoplasm. The latter also contains aggregates of ribosomes and inclusion bodies composed of ribosomes in a matrix of virus-specific protein. Progeny viral RNA and viral proteins (G1, G2, N) migrate to the cell wall. Novel virions include cellular ribosomes and are released by budding. (Authors' legend.) (From Palmer, E.L. and Martin, M.L., *Electron Microscopy in Viral Diagnosis*, CRC Press, Boca Raton, FL, 1988, 139. With permission.)

ARENAVIRIDAE

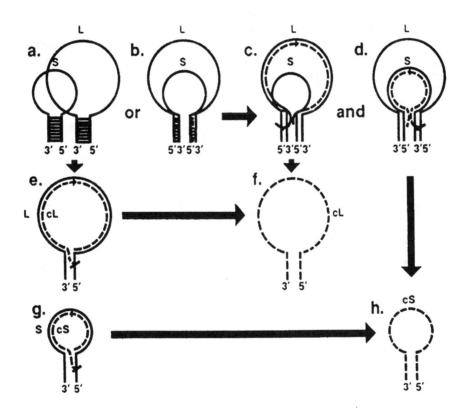

FIGURE 186

Arenavirus replication. The model assumes that replication starts at paired viral RNA termini. S and L genomic segments are shown with panhandle structures due to terminal base pairing. Replicative intermediates are represented by dashed lines. Initiation of replication from a monomolecular (self-paired) structure (a) would result in replication events (e) and (g). Initiation of replication from a bimolecular structure (b) would result in simultaneous replication of L (c) and S (d). Replication may be unprimed or primed by a fragment complementary to the 3′ end (possibly derived by cleavage of five nucleotides from the 5′ end of either L or S). The replicative intermediates cL (f) and cS (h) from monomolecular and bimolecular structures would be the same. (From Salvato, M.S., in *The Arenaviridae,* Salvato, M.S., Ed., Plenum Press, New York, 1993, 133. With permission.)

7.II.B. BUNYAVIRIDAE

Linear (–) sense ssRNA, segmented
Helical, enveloped
Vertebrates, invertebrates, plants

This large family with five genera is named after Bunyamwera, a locality in Uganda, and includes many "arboviruses" and some plant pathogens (genus *Tospovirus*). Each genus has unique replicative properties and often extremely different biological characteristics. The genera *Phlebovirus* and *Tospovirus* have ambisense RNA. Particles are enveloped, roughly spherical, and 80–120 nm in diameter. Envelopes contain three circular nucleocapsids constituted by RNA molecules circularized by protein. Nucleocapsid and envelope proteins are encoded in the S and M segments, respectively. Replication and assembly occur in the cytoplasm. Viruses generally display an unusual Golgi-associated morphogenesis. Viral glycoproteins are inserted into Golgi membranes. Viruses acquire envelopes by budding into Golgi cisternae and accumulate there, reach the cell surface within transport vesicles, and are released by exocytosis.

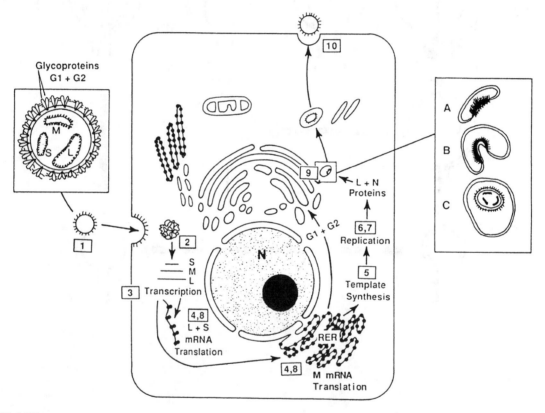

FIGURE 187

Probable replication pathway of bunyaviruses. Virions (left box) contain large (L), medium (M), and small (S) RNA molecules complexed with nucleocapsid proteins. The bilaminar viral envelope carries viral glycoproteins (G1 and G2). The replication sequence includes: (1) viral attachment, (2) uncoating, (3) primary transcription to yield viral messenger RNAs (mRNA), (4) translation of L and S segment mRNA on free ribosomes and translation of M segment mRNA on membrane-bound ribosomes, (5) synthesis of antigenome templates, (6) genome replication, (7) secondary transcription, (8) further translation, (9) terminal glycosylation of G1 and G2 and assembly of virus particles by budding into Golgi vesicles, and (10) transport of cytoplasmic vesicles to the cell surface and release of mature virus. The stages of budding into membrane vesicles (right box) are (A) accumulation of ribonucleoproteins at the portion of the cytoplasmic face of the plasma membrane which has G1 and G2 embedded into it, (B) involution of membranes, and (C) completion of budding to yield a mature virion within a cytoplasmic vacuole. N, nucleus; RER, rough endoplasmic reticulum. (From Schmaljohn, C.S. and Patterson, J.L., in *Virology,* 2nd ed., Vol. 1, Fields, B.N. and Knipe, D.M., Eds.-in-chief, Raven Press, New York, 1990, 1175. With permission.)

BUNYAVIRIDAE

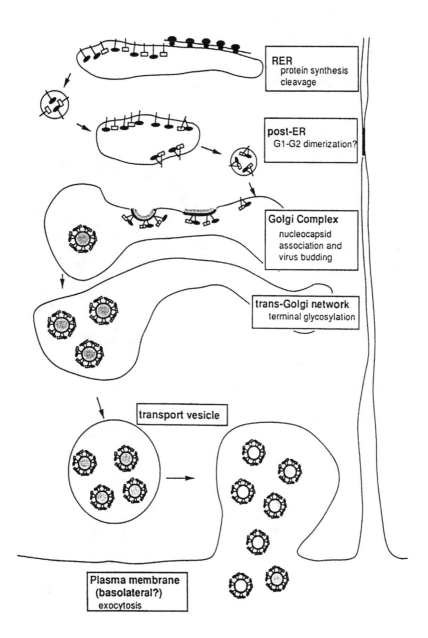

FIGURE 188

Bunyavirus glycoprotein transport and assembly. After translocation into the rough endoplasmic reticulum (RER), glycoproteins are transported to the Golgi complex. Virions are assembled by budding at the membranes of the Golgi complex, are further transported through the *trans*-Golgi network, and are released by exocytosis. In polarized epithelial cells, release appears to occur at the basolateral membranes. (From Matsuoka, Y., Chen, S.Y., and Compans, R.W., *Curr. Topics Microbiol. Immunol.*, 169, 161, 1991. ©Springer-Verlag, Berlin. With permission.)

BUNYAVIRIDAE

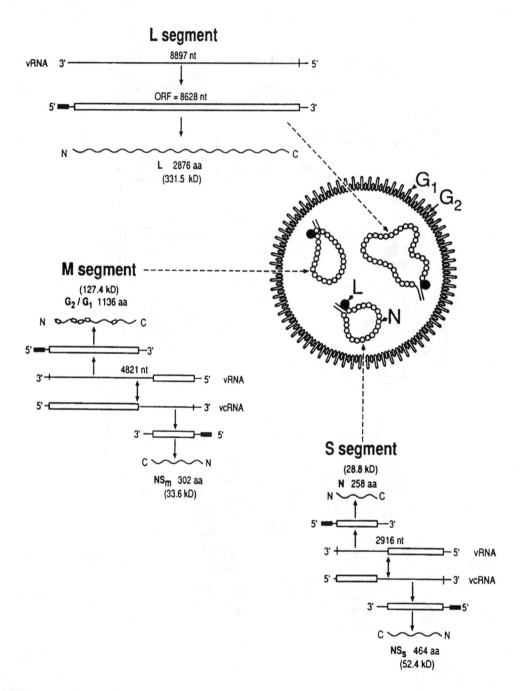

FIGURE 189

Genome structure and strategy of gene expression of tomato spotted wilt virus (*Tospovirus* genus). Within the enveloped virion, the three genomic RNAs form pseudocircular structures by base pairing of complementary terminal sequences. Open reading frames in the viral (v) and complementary (vc) strand are indicated as open bars. Messenger RNAs are provided with capped sequences (black boxes) derived from cellular mRNAs. (Reprinted from Goldbach, R. and De Haan, P., *Semin. Virol.,* 4, 381, 1993. By permission of the publisher Academic Press Ltd., London.)

BUNYAVIRIDAE

BUNYAVIRUSES

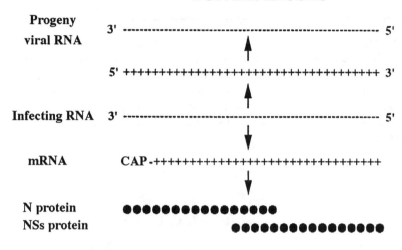

PHLEBOVIRUSES

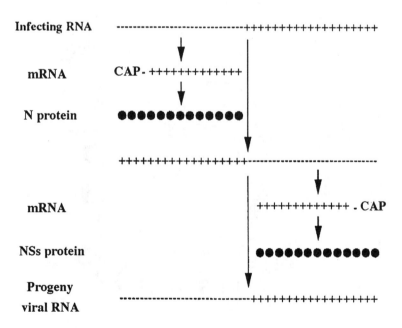

FIGURE 190

Comparative coding, transcription, and replication of the S segment of bunyaviruses and phleboviruses (ambisense, below). The S segment encodes the nucleocapsid protein N and the nonstructural protein NSs. These genes overlap partially in bunyaviruses and are transcribed into subgenomic mRNA in phleboviruses. Messenger RNAs are capped. (Redrawn by M. Tremblay after Bishop, D.H.L., *Intervirology*, 24, 79, 1985. With permission.)

7.II.C. ORTHOMYXOVIRIDAE

Linear (−) sense ssRNA, segmented
Helical, enveloped
Vertebrates, invertebrates

This family includes the human influenza viruses, has four small genera, and occurs in mammals, birds, and ticks. Particles are enveloped, usually spherical, and 80–120 nm in diameter; filamentous forms are common. The genome consists of six to eight nucleoprotein complexes (influenza A and B viruses, eight segments; influenza C, seven segments). Virions contain RNA polymerase. Cells are infected by receptor-mediated endocytosis. The nucleus is the site of RNA transcription and replication. Viral proteins are synthetized in the cytoplasm. Newly formed viral envelope glycoproteins and nucleocapsids migrate to the cell membrane and novel viruses are liberated by budding. Gene reassortment, resulting in major antigenic shifts, occurs during mixed infections by viruses of the same species.

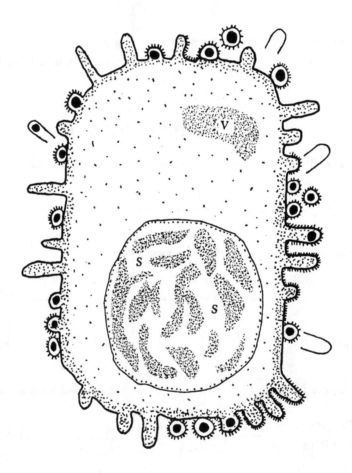

FIGURE 191

Cell showing (ortho)myxovirus replication (influenza, fowl plague). Nucleocapsids are assembled in the nucleus (S or "soluble antigen"). Hemagglutinin antigen (V) is found in the cytoplasm and moves to the cytoplasmic membrane to form peplomers of the viral envelope. Nucleocapsids then move to positions beneath the altered cytoplasmic membrane and complete viruses are released by budding. (From Bernhard, W., in *Ciba Foundation Symposium on Cellular Injury,* De Reuck, A.V.S. and Knight, J., Eds., J. & A. Churchill, London, 1964, 209. With permission.)

ORTHOMYXOVIRIDAE

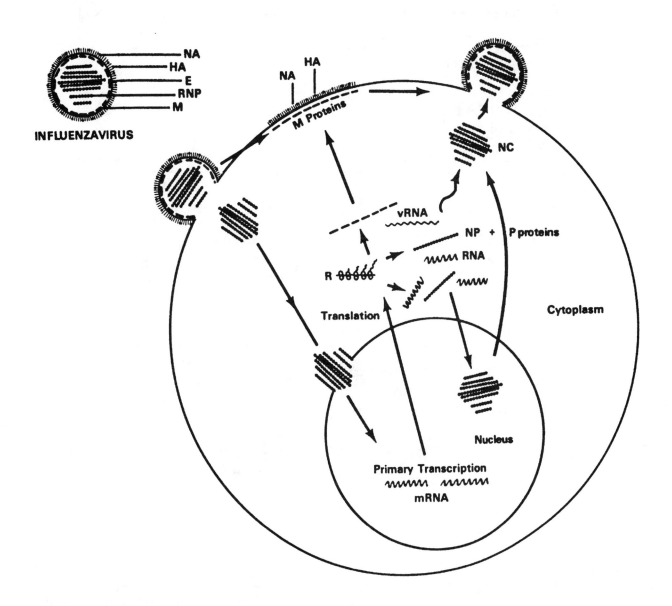

FIGURE 192

Replication of influenza virus. A complete virion is shown at the left. Viruses are formed at the plasma membrane without evidence of a cytoplasmic precursor. Primary transcription occurs in the nucleus. E, envelope; HA, hemagglutinin; M, matrix protein; NA, neuraminidase; NC, nucleocapsid; NP, nucleoprotein; R, ribosomes. (From Palmer, E.B. and Martin, M.L., *Electron Microscopy in Viral Diagnosis,* CRC Press, Boca Raton, FL, 1988, 108. With permission.)

ORTHOMYXOVIRIDAE

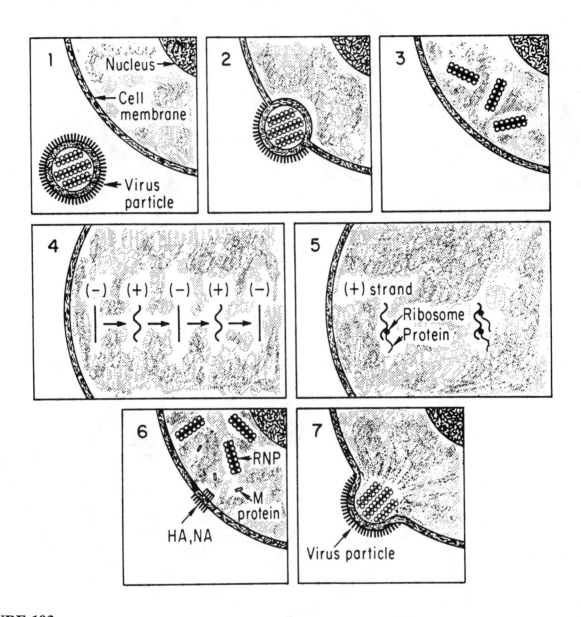

FIGURE 193

Another view of influenza virus replication (only three ribonucleic complexes are shown). (1) Adsorption: virus attaches to cell receptors via the hemagglutinin. (2) Penetration: engulfment of virus particle or fusion with cell membrane. (3) Release of ribonucleic complex into cytoplasm. (4) RNA synthesis: incoming viral (–) RNA is transcribed by virus-associated polymerase to make complementary (+) RNA. This is used as a message and template for further virus-specific RNA synthesis (+ and –). (5) Protein synthesis: takes place in the cytoplasm, but P, NP, and NS proteins can be found in the nucleus; glycosylation occurs in rough and smooth membranes. (6) Assembly: RNAs incorporated into a ribonucleoprotein complex (RNP); hemagglutinin (HA) and neuraminidase (NA) are inserted into the cell membrane; the matrix (M) protein probably associates with the inner surface of the cell membrane. (7) Release: components assemble at the cell membrane and viruses bud into the extracellular space while neuraminidase prevents auto-agglutination. (From Palese, P. and Ritchey, M.B., in *Virology and Rickettsiology,* Vol. I, Part 1, Hsiung, G.-D. and Green, R.H., Eds., CRC Press, Boca Raton, FL, 1978, 337. With permission.)

ORTHOMYXOVIRIDAE

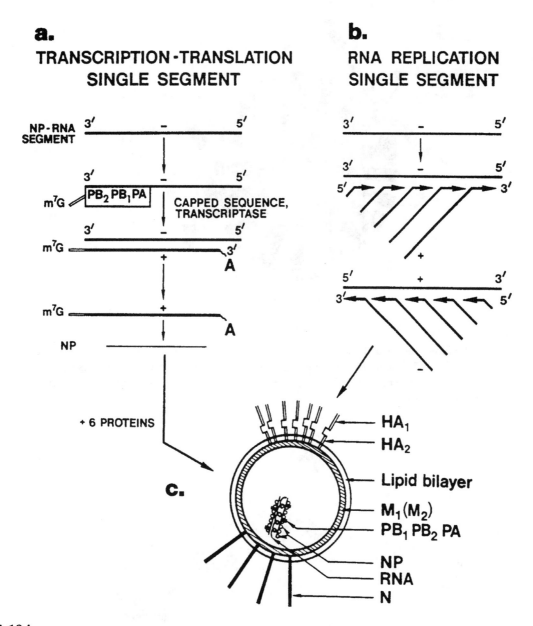

a.
TRANSCRIPTION-TRANSLATION
SINGLE SEGMENT

b.
RNA REPLICATION
SINGLE SEGMENT

FIGURE 194

Replication of influenza virus. (a) Transcription and translation of the RNA segment for nucleocapsid protein (NP) serves as an example for the other segments. A capped primer (open rectangle) donated by cell RNA associates with RNA polymerase, PB_2, PB_1, PA. This complex initiates mRNA synthesis using the genomic RNA as a template to produce NP mRNA which is shorter than the full-length template RNA. All but the 5′ and 3′ terminal regions of viral mRNA are translated into NP. (b) Two replicative intermediates are involved in replication of each RNA segment, one with (–) strand RNA serving as a template and the other with a (+) strand RNA template. Viral nucleocapsids are formed by association of the PB_2, PB_1, PA complex and NP with the eight segments of RNA. Nucleocapsids comprising all eight RNA segments bud through areas of the cell membrane which have HA and N proteins on their external surfaces and M located at their cytoplasmic sides to form viral particles (c, only one nucleocapsid shown). HA is illustrated in its cleaved form; HA_2 and HA_1 are linked by a disulfide bond. HA and N are actually interspersed on the virus particle and HA spikes are more abundant. (From Fenger, T.W., in *Textbook of Human Virology,* 2nd ed., Belshe, R.B., Ed., Mosby-Year Book, St. Louis, 1991, 74. With permission.)

ORTHOMYXOVIRIDAE

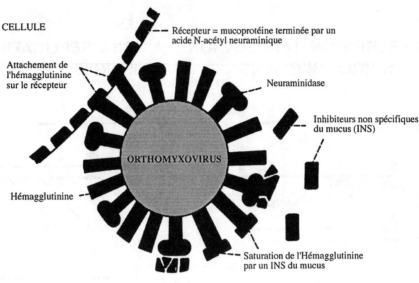

FIGURE 195

The first step of multiplication. The virus attaches to the receptor, a mucoprotein with a terminal N-acetyl muramic acid molecule, via the hemagglutinin. The mucus of the extracellular space contains nonspecific inhibitors (INS) which may saturate the hemagglutinin, but are destroyed by neuraminidase. (From Huraux, J.M., Nicolas, J.C., and Agut, H., *Virologie,* Flammarion, Paris, 1985, 168. With permission.)

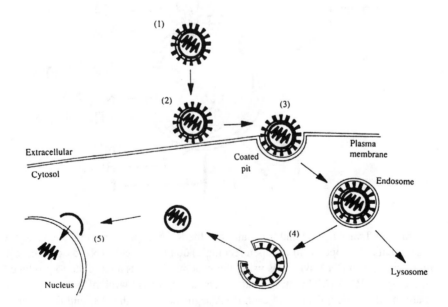

FIGURE 196

Pathway of influenza virus entry. (1) Free virus. (2) Virus adsorbs to cell membrane. (3) Virus is taken up into a vesicle by endocytosis. (4) Virus is uncoated whereby the virus genome becomes functional. A proton pump in the vesicle membrane acidifies the content of the vesicle, causing fusion of viral and vesicle membranes and entry of the nucleocapsid into the cytoplasm. (5) The nucleocapsid is transported into the cell nucleus. (From Outlaw, M.C. and Dimmock, N.J., *Epidemiol. Infect.,* 106, 205, 1991. Reprinted with the permission of Cambridge University Press.)

ORTHOMYXOVIRIDAE

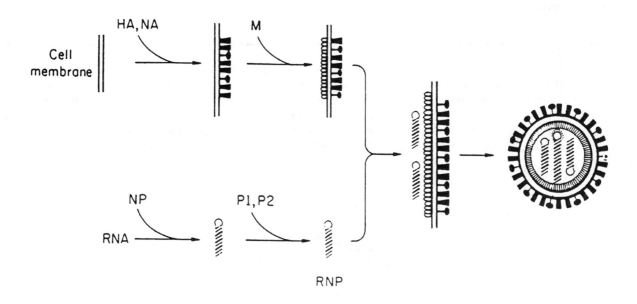

FIGURE 197

Assembly pathway of influenza virus. For abbreviations, see legend to Fig. 192. (From Compans, R.W., Meier-Ewert, H., and Palese, P., *J. Supermol. Struct.*, 2, 496, 1974. ©1974 John Wiley & Sons. Reprinted by permission of Wiley-Liss, Inc., a subsidiary of John Wiley & Sons, Inc.)

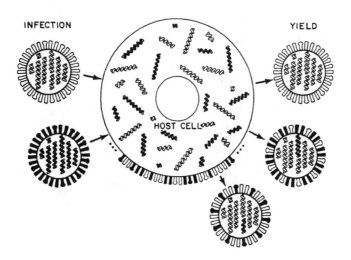

FIGURE 198

Intratypic genetic reassortment of influenza viruses. Simultaneous infection of a cell with any pair of influenza A viruses has the potential for yielding reassortment viruses of mixed genotype. The diagram illustrates 3 of 256 possible reassortant progeny from infection with 2 parental viruses. (From Kilbourne, E.D., *Influenza*, Plenum Medical Books, New York, 1987, 127. With permission.)

7.II.D. PARAMYXOVIRIDAE

Linear (–) sense ssRNA, nonsegmented
Enveloped, helical
Vertebrates

Viruses are mammal-specific and divided into two subfamilies (*Paramyxovirinae, Pneumovirinae*) and four genera. They include important human and animal pathogens such as measles, mumps, and rinderpestvirus. Particles are roughly spherical and 150–300 nm in size; filamentous forms are common as in orthomyxoviruses. The envelope contains RNA polymerase and a relatively rigid nucleocapsid with a single RNA molecule. Viruses enter cells by membrane fusion. Viral syntheses take place in the cytoplasm. The genome is transcribed into six to ten capped mRNAs, each of them encoding one or two proteins. Viral glycoproteins are inserted at the plasma membrane. Nucleocapsids assemble and exit there by budding, acquiring an envelope in the process.

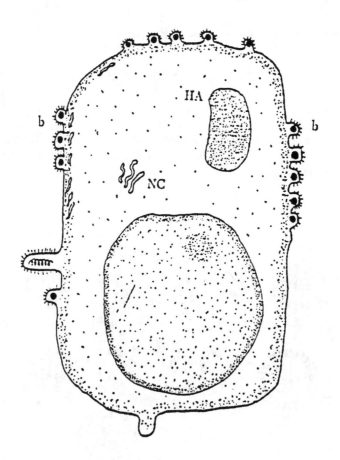

FIGURE 199

Cell showing paramyxovirus (Newcastle disease virus) replication. Both nucleocapsid (NC) and hemagglutinin (HA) antigen are synthetized and assembled in the cytoplasm. Hemagglutinin antigen moves to the cytoplasmic membrane to become envelope peplomers. Nucleocapsids move to a position beneath the cytoplasmic membrane and then become enveloped by budding (b). (From Fenner, F., *The Biology of Animal Viruses*, Vol. I, Academic Press, New York, 1968, 264. With permission.)

PARAMYXOVIRIDAE

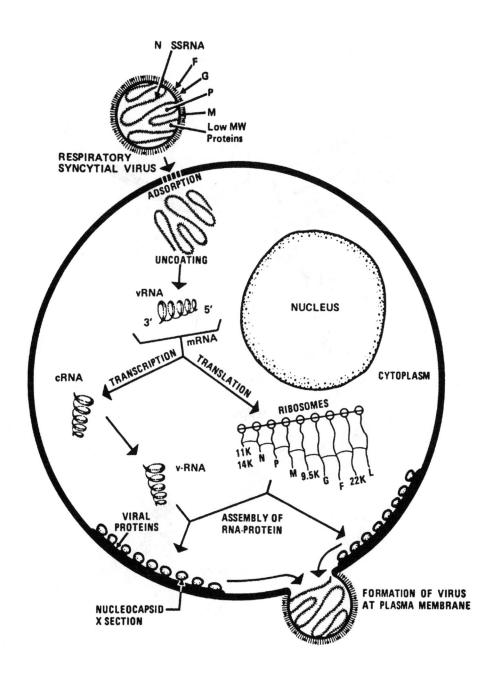

FIGURE 200

Replication of respiratory syncytial virus. Proteins F, G, M, N, and P are constitutive viral proteins. Nucleocapsids are assembled just under the plasma membrane and complete virions exit by budding. (From Palmer, E.L. and Martin, M.L., *Electron Microscopy in Viral Diagnosis*, CRC Press, Boca Raton, FL, 1988, 114. With permission.)

PARAMYXOVIRIDAE

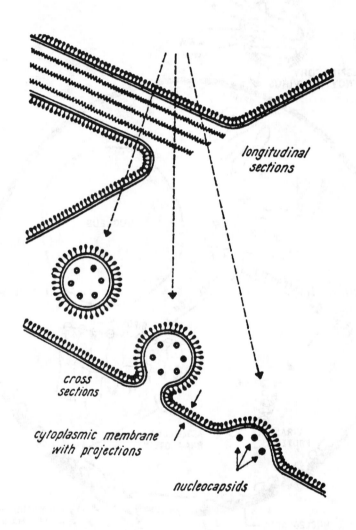

FIGURE 201

Paramyxoviruses produce round and filamentous forms. Ultrathin sections of infected cells can be misleading with respect to their relative amounts because cross-sections of filamentous forms at different angles could produce most if not all of the round structures seen. (From Berthiaume, L., Joncas, J., and Pavilanis, V., *Arch. Ges. Virusforsch.,* 45, 39, 1974. With permission.)

PARAMYXOVIRIDAE

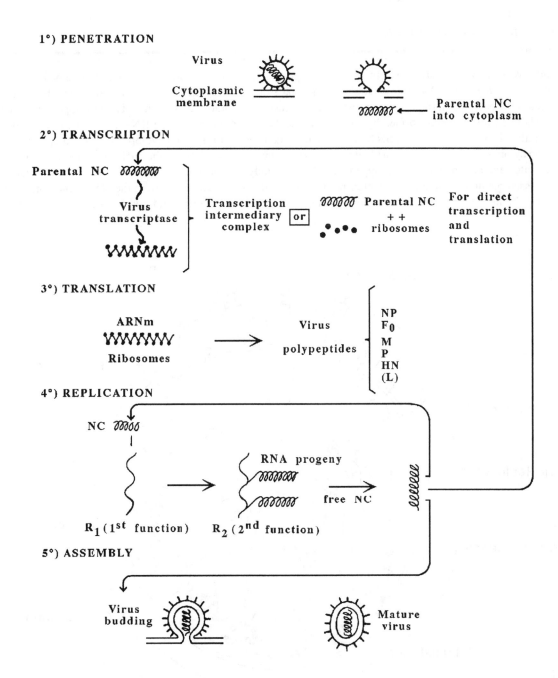

FIGURE 202

Paramyxovirus replication. R_1 represents the first activity of RNA polymerase, the production of complementary mRNA. A replicative intermediate is formed and the mRNA produces progeny genomes via R_2, the second function of the polymerase. Free nucleocapsids (NC) may exit by budding or be recycled for amplification of R_1 or the pool of transcription intermediates. (Adapted from Georges, J.-C., in *Virologie Médicale,* Maurin, J., Ed., Flammarion, Paris, 1985, 475. With permission.)

7.II.E. RHABDOVIRIDAE

Linear (–) sense ssRNA, nonsegmented
Helical, enveloped
Vertebrates, invertebrates, plants

This large and diversified family has an exceptional host range and occurs in vertebrates, insects, and plants. It comprises five genera, notably *Lyssavirus* (rabies) and *Vesiculovirus* (vesicular stomatitis). Particles are enveloped, bullet-shaped (vertebrate and insect viruses) or bacilliform (many plant viruses), and measure $100–430 \times 50–100$ nm. The envelope contains a tubular nucleocapsid about 50 nm in width, formed by a coiled filament of protein and a single RNA molecule. RNA polymerase is associated with the virion. Viruses enter cells by receptor-mediated endocytosis. The envelope is removed and viral syntheses take place in the cytoplasm; however, certain plant rhabdoviruses replicate their RNA in the nucleus. Genes are transcribed from template RNA into generally monocistronic capped mRNAs. Assembly sites depend on virus and host. In animal rhabdoviruses, nucleocapsids assemble in the cytoplasm and, after insertion of viral glycoproteins, bud from the plasma membrane (also from cytoplasmic membranes). Certain plant rhabdoviruses are assembled in the nucleus and bud from the nuclear membrane.

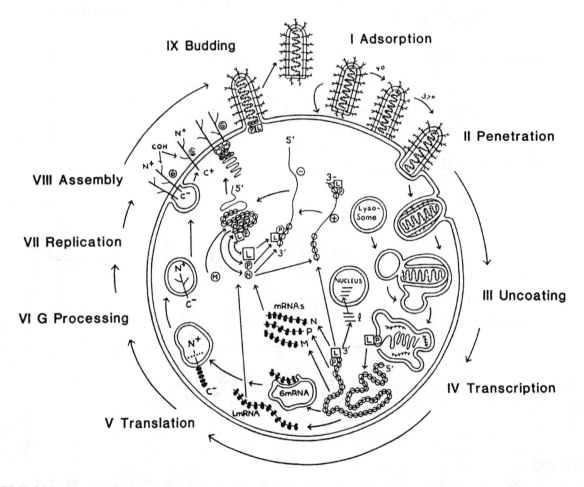

FIGURE 203

Replication cycle of vesicular stomatitis virus (VSV). Events such as transcription and translation as well as genome replication and translation proceed simultaneously. (From Wagner, R.R., in *The Rhabdoviridae*, Wagner, R.R., Ed., Plenum Press, New York, 1987, 9. With permission.)

RHABDOVIRIDAE

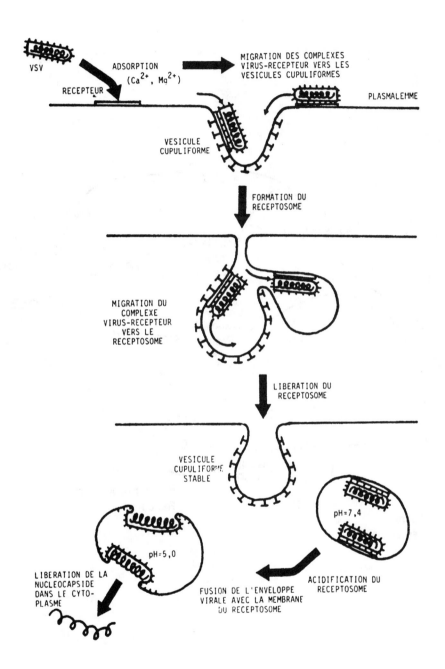

FIGURE 204

VSV adsorption and penetration. After adsorption, virus-receptor complexes migrate toward cup-like invaginations and then to "receptosomes." Invagination and receptosome separate again. The former remains at the cell surface and the latter in the cytoplasm. Its content is acidified, its membrane fuses with the viral envelope, and the viral nucleocapsid is released into the cell. (From Danglot, C., in *Virologie Médicale,* Maurin, J., Ed., Flammarion, Paris, 1985, 57. With permission.)

RHABDOVIRIDAE

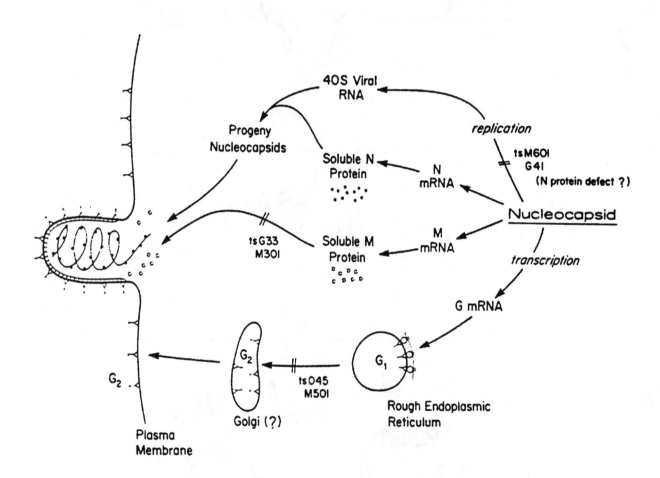

FIGURE 205

Maturation of major structural proteins of VSV and proposed blocking sites in virus assembly for certain temperature-sensitive (ts) mutants. Protein G is inserted into the endoplasmic reticulum and partially glycosylated. Final glycosylation occurs in light-density membranes; soon after, G appears at the cell surface. Protein M, at first soluble, is progressively incorporated into membranous structures with the density of virions and then quickly appears in extracellular virions. The nucleocapsid (N) protein is also soluble and is later incorporated into nucleocapsids that attach to the cell membrane prior to budding. The movement of G, M, and N proteins to the cell surface appears to be interdependent. (From Knipe, D.M., Baltimore, D., and Lodish, H.F., *J. Virol.,* 21, 1149, 1977. With permission.)

RHABDOVIRIDAE

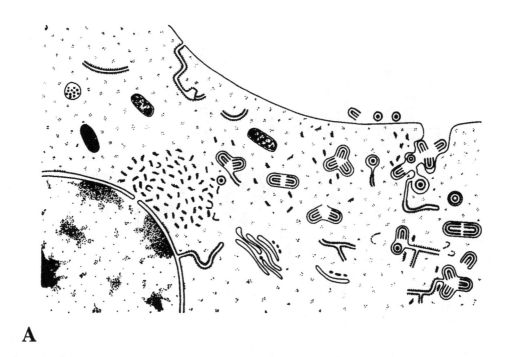

A

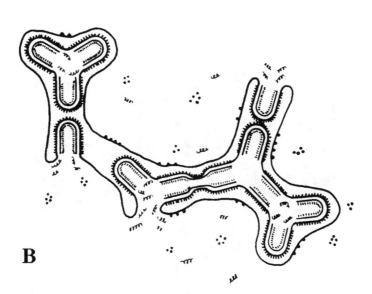

B

FIGURE 206

Morphogenesis of lyssaviruses. (A) General view of terminal stages. (B) Last events of a complex envelope formation in cytoplasmic cisterns. The complex encompasses three triplets. (Adapted from Gosztonyi, G., *Curr. Topics Microbiol. Immunol.*, 187, 43, 1994. © Springer-Verlag, Berlin. With permission.)

RHABDOVIRIDAE

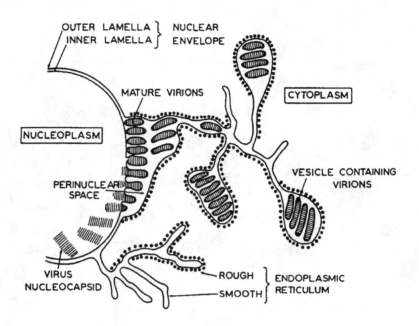

FIGURE 207

Possible sequence of events in synthesis of a plant rhabdovirus. Nucleocapsids are synthetized in the nucleoplasm and extruded into the perinuclear space while acquiring viral envelopes. Viral particles can then migrate into the endoplasmic reticulum to form aggregates enclosed in vesicles. (From Francki, R.I.B., *Adv. Virus Res.,* 18, 257, 1973. With permission.)

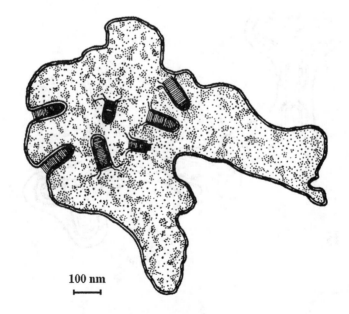

100 nm

FIGURE 208

Formation of bullet-shaped particles from membrane material in a plant cell infected by broccoli necrotic yellows virus. Both complete and partially assembled particles are observed in association with cell plasmalemma and endoplasmic reticulum. (From Horne, R.W., *Virus Structure,* Academic Press, New York, 1974, 35. With permission.)

RHABDOVIRIDAE

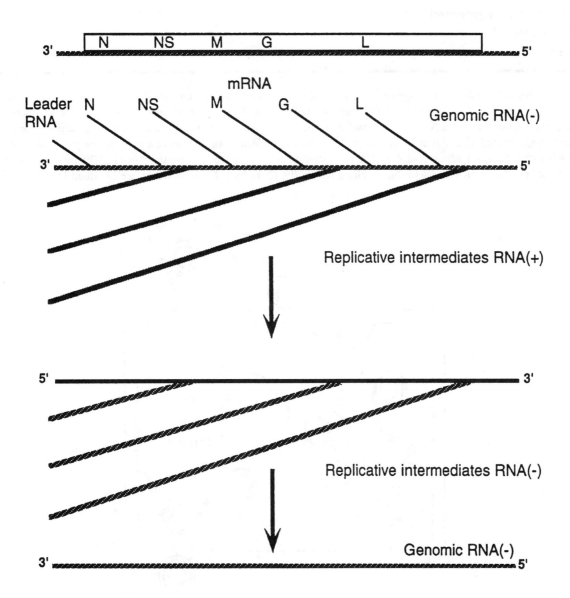

FIGURE 209

Replication of VSV. The upper line shows the gene order of VSV. Thick lines are replicative intermediate (or genome) RNA-N protein complexes; thin lines are leader RNA, or mRNAs. (From Wunner, W.H. et al., in *Virus Taxonomy, Sixth Report of the International Committee on Taxonomy of Viruses,* Murphy, F.A. et al., Eds., Springer, Vienna, *Arch. Virol.,* Suppl. 10, 185, 1995. © Springer-Verlag. With permission.)

III. dsRNA Viruses

7.III.A. CYSTOVIRIDAE

Linear dsRNA, segmented
Cubic, enveloped
Bacteria

The only member is *Pseudomonas* phage φ6, an enveloped virus about 75 nm in diameter. The envelope contains an icosahedral capsid with three RNA molecules and RNA polymerase. Virions adsorb to pili, are transported to the cell wall, digest it locally, and free capsids enter the cell. RNA transcription/replication is semi-conservative. (+) strands are transcribed from viral RNA within the capsid, extruded, and translated into viral proteins. A dodecahedral polymerase complex or procapsid is formed and packages (+) strands, from which complementary (–) strands are synthetized within the particle. The procapsid is completed by an external shell and provided with an envelope within the bacterium. Mature viruses are liberated by cell lysis.

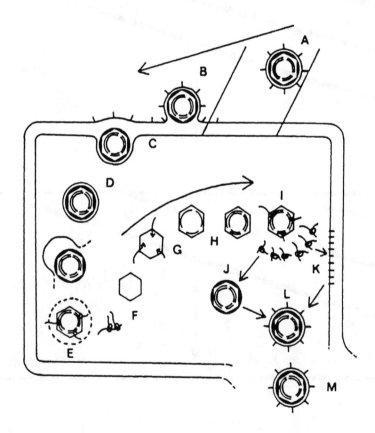

FIGURE 210

Events of the φ6 life cycle. (A) Adsorption to host pilus. (B) Fusion of viral envelope with bacterial outer membrane. (C) Local digestion of cell wall peptidoglycan. (D) Penetration of nucleocapsid through cytoplasmic membrane. (E) Nucleocapsid uncoating and early transcription. (F). Synthesis of early proteins and assembly of procapsids. (G) Packaging of viral ssRNAs. (H) dsDNA synthesis. (I) Late transcription. (J) Synthesis and assembly of nucleocapsid shell protein. (K) Synthesis of viral envelope proteins on the cytoplasmic membrane. (L) Intracellular translocation of envelopes on the nucleocapsids. (M) Release of progeny phage by cell lysis. (From Olkkonen, V., Ph.D. thesis, Department of Genetics, University of Helsinki, 1991; also shown in Ref. 191. With permission.)

CYSTOVIRIDAE

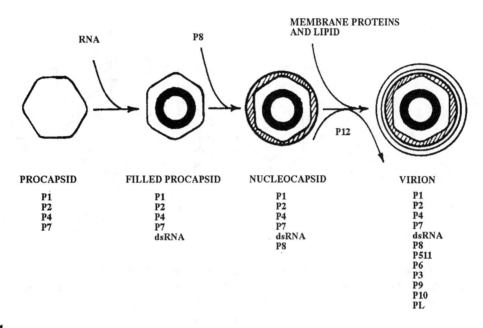

FIGURE 211

Morphogenetic pathway of φ6. Procapsids are assembled and mature to nucleocapsids by acquisition of dsRNA and protein P8. Nucleocapsids acquire an envelope comprising phospholipid (PL) and virus-specified proteins. (From Bamford, D.H. and Mindich, L., *Virology,* 107, 222, 1980. With permission.)

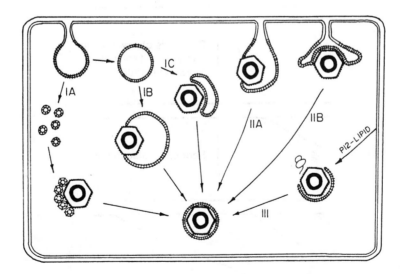

FIGURE 212

Models for membrane acquisition by bacteriophage φ6. Short lines perpendicular to membranes represent phage membrane proteins. (IA) Phage proteins and host lipids are transferred from the host membrane in small vesicles. (IB) A large free vesicle of normal orientation is entered by a nucleocapsid or (IC) everts over a nucleocapsid. (IIA) A nucleocapsid enters an attached vesicle of normal orientation or (IIB) is covered by an everting, reversed-orientation, attached vesicle. (III) The viral membrane is assembled on the surface of the nucleocapsid. (From Stitt, B.L. and Mindich, L., *Virology,* 127, 446, 1983. With permission.)

CYSTOVIRIDAE

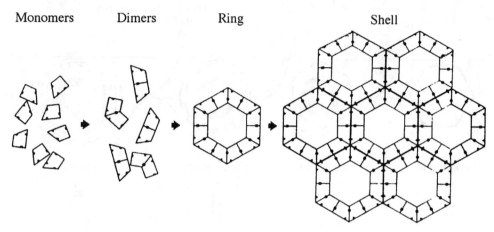

FIGURE 213

Assembly of the outer shell of φ6 nucleocapsid. Asymmetric P8 monomers of arbitrary shape dimerize in the presence of Ca⁺⁺. The dimers form closed rings which, in turn, assemble hexagonally into the outer shell. As drawn, there are two alternative dimers, straight and angled, and three distinct binding sites are required (dots, squares, diamonds). (Adapted from Ktistakis, N.T., Kao, C.-Y., and Lang, D., *Virology,* 166, 91, 1988. With permission.)

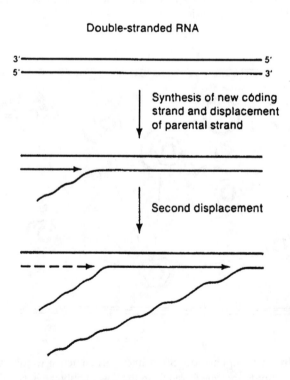

FIGURE 214

Synthesis of mRNA in φ6 by displacement of a parental RNA strand. (From Freifelder, D., *Molecular Biology, An Introduction to Prokaryotes and Eukaryotes,* Science Books International, Boston, 1983, 662. ©1983 Boston: Jones and Bartlett Publishers. Reprinted with permission.)

7.III.B. PARTITIVIRIDAE

Linear dsRNA, segmented
Cubic, naked
Plants, fungi

This small family includes two genera of plant viruses and two genera of fungal viruses. Particles are isometric, 30–40 nm in diameter, and contain two (monocistronic?) RNA molecules and RNA polymerase. Transcription/replication occurs in the cytoplasm and is semi-conservative. Plant viruses are transmitted via the seed embryo and fungal viruses are transmitted intracellularly.

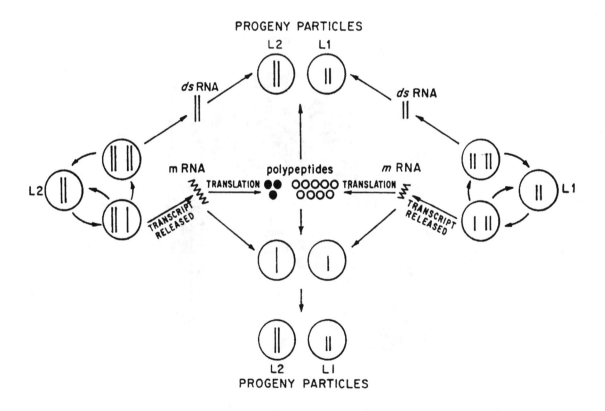

FIGURE 215

Replication of *Penicillium stoloniferum* virus S (PsV-S). Open circles, capsid protein subunits; closed circles, RNA polymerase; solid lines, genomic or replicating RNA strands; wavy lines, mRNA. (Reprinted from Buck, K.W., in *Viruses and Plasmids of Fungi,* Lemke, P.A., Ed., Marcel Dekker, New York , 1979, 93. By courtesy of Marcel Dekker, Inc.)

7.III.C. REOVIRIDAE

Linear dsRNA, segmented
Cubic, naked
Vertebrates, invertebrates, plants

This large family has nine genera. Members are found in mammals, birds, fish, shellfish, insects, ticks, crustaceans, and plants. Reovirus genera differ in host range, capsid structure, number of RNA molecules, and modes of virus entry and egress. Members of the genera *Orbivirus* and *Coltivirus* multiply in arthropods and vertebrates, whereas plant reoviruses multiply in insects and plants. Members of the genus *Cypovirus,* which infect insects and crustaceans, are included into polyhedral protein crystals. Particles are icosahedra 60–80 nm in diameter. Capsids consist of one or two outer coats and an inner core containing 10–12 RNA molecules and RNA polymerase. Most viral RNA molecules are monocistronic, each segment coding for a different protein. Transcription/replication occurs in the cytoplasm and is conservative. Transcription of genomic RNA to mRNA starts in internalized viral cores. Novel mRNAs are extruded through the apices of the cores. It is believed that mRNA molecules and viral proteins form procapsids and that the mRNAs are transcribed there into complementary (–) strands with which they form dsRNA. Viruses often assemble within inclusion bodies.

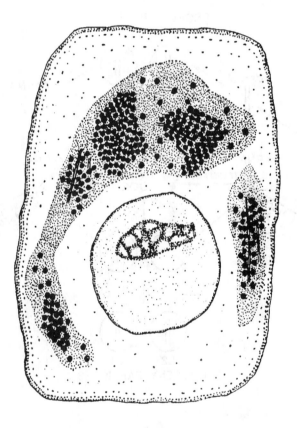

FIGURE 216
Cell showing reovirus replication. Large crystalline arrays of virions are found in perinuclear inclusion bodies or "factories." (From Bernhard, W., in *Ciba Foundation Symposium on Cellular Injury,* De Reuck, A.V.S. and Knight, J. Eds., J. & A. Churchill, London, 1964, 209. With permission.)

REOVIRIDAE

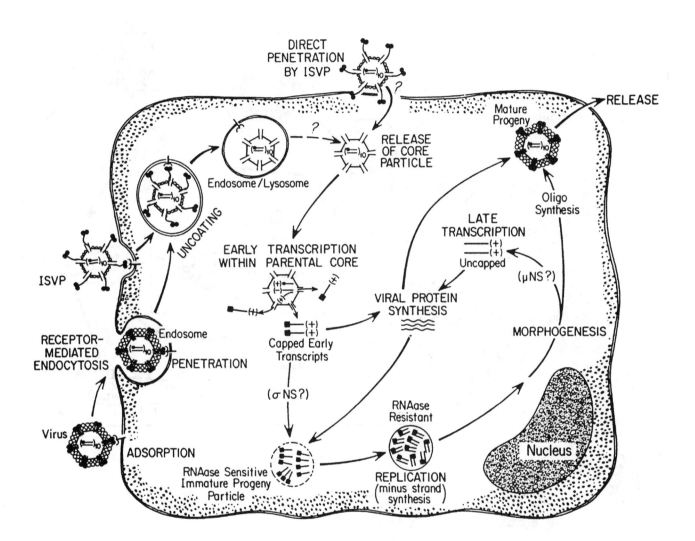

FIGURE 217

Reovirus replication cycle (drawing by Pamela Battaglino). Viruses enter cells by receptor-mediated endocytosis and are partially uncoated within lysosomes by digestion of the outer capsid. Core particles are released into the cytoplasm and the genome is partially transcribed within core particles. The RNA is replicated within nascent progeny subviral particles. Structural proteins assemble to form progeny cores and outer capsids. Mature viruses are released by cell lysis. (Authors' legend.) (From Schiff, L.A. and Fields, B.N., in *Virology,* 2nd ed., Vol. 2, Fields, B.N. and Knipe, D.M., Eds.-in-chief, Raven Press, New York, 1990, 1275. With permission.)

REOVIRIDAE

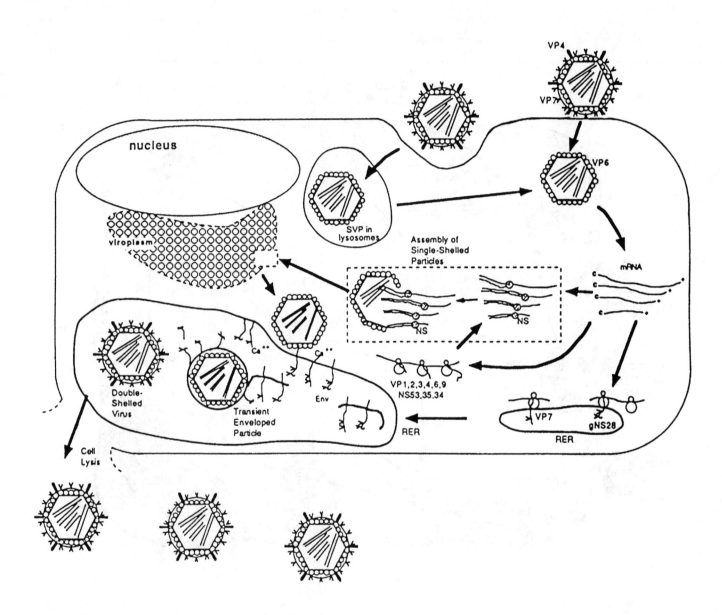

FIGURE 218

Major features of the rotavirus cycle. Many details are inferred from reovirus replication. Viruses enter by receptor-mediated endocytosis or direct penetration through the plasma membrane. Early transcription produces single-shelled subviral particles. These assemble in cytoplasmic viroplasms, bud through membranes of the endoplasmic reticulum, and acquire envelopes in the process. The envelopes are replaced by the outer capsid of mature viruses, which are released by cell lysis. (Authors' legend.) (From Estes, M.K., in *Virology,* 2nd ed., Vol. 2, Fields, B.N. and Knipe, D.M., Eds.-in-chief, Raven Press, New York, 1990, 1329. With permission.)

REOVIRIDAE

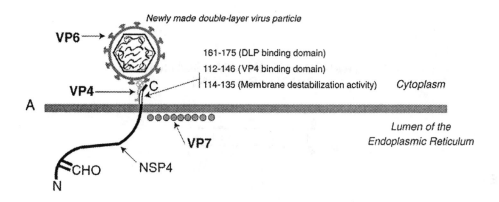

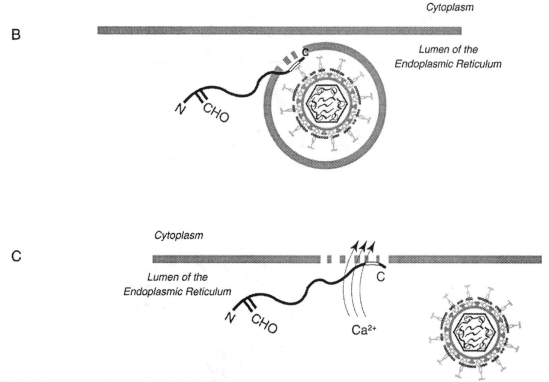

FIGURE 219

Rotavirus maturation. During a unique process, rotaviruses obtain a transient envelope when newly synthetized immature particles bud into the endoplasmic reticulum (ER). This membrane is lost again and a layer of glycoprotein VP7 forms the outer shell of the virion. The nonstructural glycoprotein NSP4 functions as an intracellular receptor in the ER membrane. (A) Newly synthetized double-layered particles (DLP) in the cytoplasm bind via VP4 to the C terminus of NSP4 and bud into the lumen of the ER. (B) VP7 undergoes a rearrangement and relocates to the interior of the enveloped particle during the budding process. VP7 interacts with VP4 inside the enveloped particles and forms the outer shell of triple-layered particles. This releases the membrane-destabilizing domain of NSP4. (C) This domain removes the envelope. CHO, glycosylation sites on NSP4; N and C, N and C termini of NSP4, respectively. (Redrawn from Tian, P., Ball, J.M., Zeng, C.Q.Y., and Estes, M.K., *J. Virol.,* 70, 6973, 1996. With permission.)

REOVIRIDAE

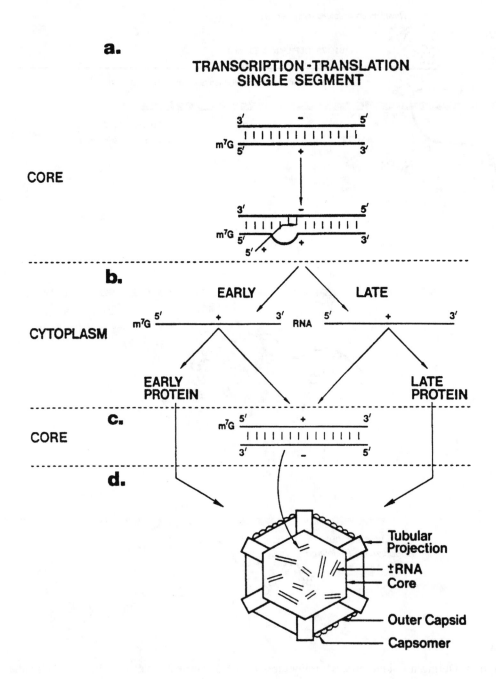

FIGURE 220

Reovirus replication. RNA transcription from each of the ten segments occurs within viral cores. (a) Transcription from a single segment. (b) Single-stranded (+) RNAs leave viral cores and are translated within the cytoplasm into early viral proteins. (c) Single-stranded (+) RNAs are also partially encapsidated to form cores in which complementary strands are synthetized. (d) Progeny cores mature to virus particles or (a) serve as sites for late mRNA synthesis. Late mRNA, unlike early mRNA, is not capped (b). (From Fenger, T.W., in *Textbook of Human Virology,* 2nd ed., Belshe, R.B., Ed., Mosby-Year Book, St. Louis, 1991, 74. With permission.)

REOVIRIDAE

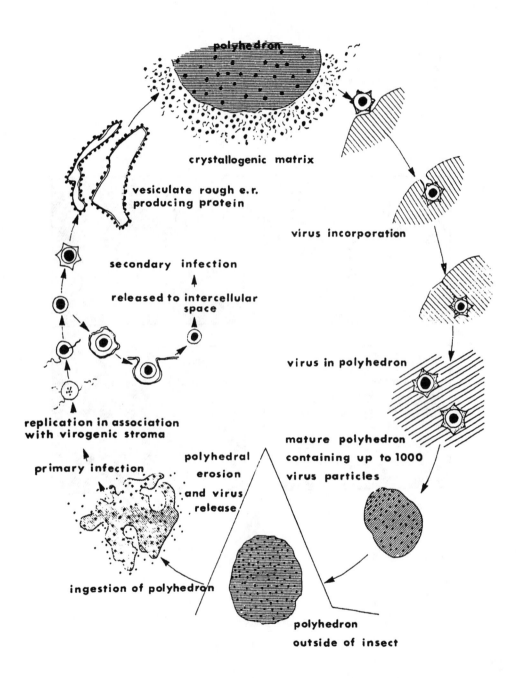

FIGURE 221

Replication of a cytoplasmic polyhedrosis virus in the larva of the monarch butterfly. Various stages of the cycle are not drawn to scale. (From Arnott, H.J., Smith, K.M., and Fullilove, S.L., *J. Ultrastruct. Res.,* 24, 479, 1968. With permission.)

REOVIRIDAE

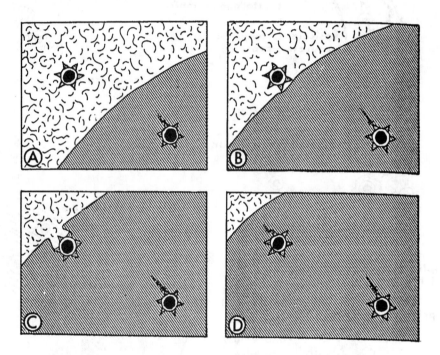

FIGURE 222

Incorporation of a cytoplasmic polyhedrosis virus into a polyhedron as seen in the larva of the monarch butterfly. (A) One particle is occluded and a second one is found near the surface of the crystal within the crystallogenic matrix. (B) Second particle at the surface of the polyhedron. (C) Second particle partly incorporated into the polyhedron. (D) Complete occlusion of second particle and formation of a false tail. (From Arnott, H.J., Smith, K.M., and Fullilove, S.L., *J. Ultrastruct. Res.*, 24, 479, 1968. With permission.)

7.III.D. TOTIVIRIDAE

Linear dsDNA, nonsegmented
Cubic, naked
Fungi, protozoa

This family includes one genus of fungal and two genera of protozoal viruses. Particles are isometric, 30–40 nm in diameter, and contain a single RNA molecule and RNA polymerase. Replication/transcription occurs in the cytoplasm and is semiconservative. (+) strand transcripts are extruded from internalized viral particles, serve as mRNA to make viral proteins, and are packaged into novel viruses. A complementary (–) strand is synthetized inside these particles to yield dsRNA. The whole process resembles the strategy of cystovirus φ6, except that no procapsid is formed.

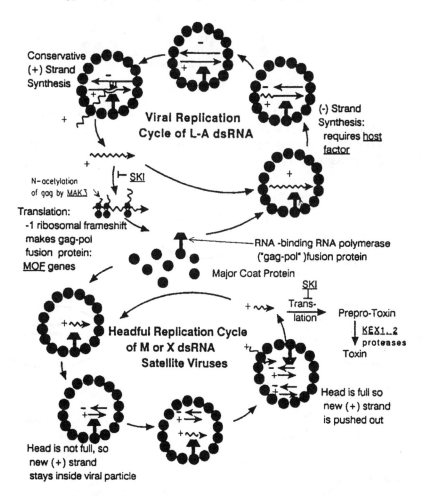

FIGURE 223

Replication cycles of yeast L-A virus (top), its deletion mutant X, and the associated satellite dsRNA M (bottom). Both (+) and (–) strand syntheses take place within virions. (+) strands are synthetized conservatively from a dsRNA template and newly synthetized (+) strands are extruded from the particles. Released (+) strands serve as mRNA and may be packaged into progeny virions or serve as templates for (–) strand RNA synthesis. Both M and X dsRNA are less than half the size of L-A virus genomic RNA and depend on it for viral proteins. As in L-A, the (+) strands of X or M dsRNAs are packaged and replicated. Because particles are designed to accommodate one L-A RNA molecule per particle, newly synthetized (+) strand RNAs are not extruded. As a result, retained (+) strands are copied to form a second dsRNA molecule within the same particle. The process continues until the particle is full ("headful replication"). (Reprinted from Bamford, D.H. and Wickner, R.B., *Semin. Virol.,* 5, 61, 1994. By permission of the publisher Academic Press Ltd., London.)

TOTIVIRIDAE

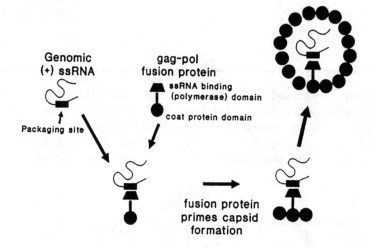

FIGURE 224

RNA packaging and assembly in virus L-A. The pol domain of the gag–pol fusion protein binds to RNA packaging sites on viral (+) strands, and the gag domain associates with free gag protein to prime polymerization of the coat. This results in the encapsidation of a single viral (+) strand, which is then converted within the particle to dsRNA. (From Wickner, R.B., *J. Biol. Chem.*, 268, 3797, 1993. With permission.)

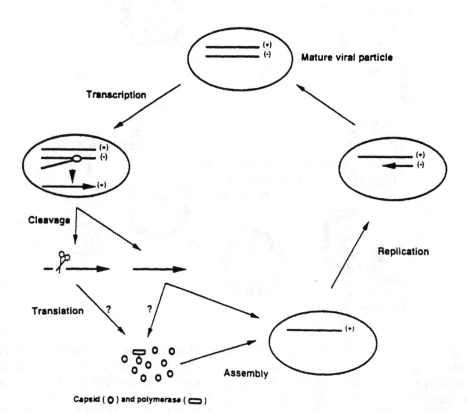

FIGURE 225

Replication of *Leishmania* RNA virus 1 (LRV1). Cleavage of viral (+) strands is a prerequisite to synthesis of viral capsid proteins, RNA polymerase, and RNA. During RNA replication, (−) strands are synthetized on a (+) strand. (Authors' legend.) (From MacBeth, K.J. and Patterson, J.L., *J. Virol.*, 69, 3458, 1995. With permission.)

Chapter 8

MISCELLANY

8.A. VIRUS-LIKE PARTICLES

Noncultivated particles resembling viruses have been observed in sectioned tissues and cells, body and culture fluids, crushed insects, plant sap, bacterial lysates, and plain water. Their viral nature is often conjectural. Part of them probably represent defective viruses; others may represent viruses awaiting discovery. The possible transmission mode of a virus-like particle of an amoeba is depicted below. It differs from the mode of infection of other viruses and suggests that many new viruses may be discovered in protozoa.

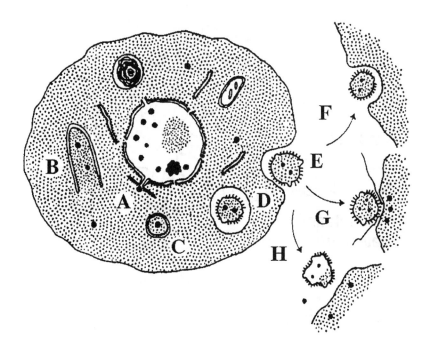

FIGURE 226

Development and possible modes of release and transfer of virus-like particles (VLPs) of *Naegleria gruberi* in cell cultures. (A) Particles produced in the nucleus exit through tubes in the nuclear envelope. (B) VLPs induce the *de novo* formation of membranous structures in the cytoplasm. (C) Structures round up, separate from the cytoplasm, and (D) become spheres within vacuoles. (E) Spheres, which have a characteristic tubular fringe, are released into the medium. They are probably vehicles for VLP transmission which might be ingested by other amoebae into food vacuoles (F) or, through contact, might allow for passage of VLPs into other cells (G). A third possibility is that VLPs are released from spheres and taken up directly by other amoebae (H). (From Schuster, F.L. and Dunnebacke, T.H., *Cytobiologie*, 14, 131, 1976. With permission.)

8.B. AGENTS WITH CIRCULAR ssRNA

Circular ssRNA
Particulate or not
Vertebrates, plants

This group of agents with possible phylogenic relationships includes the hepatitis delta virus, viroids, and circular satellite RNAs. They share circular ssRNA genomes, small genome size, absence of a capsid, absence of mRNA functions in the genome, and replication via a rolling-circle mechanism. Hepatitis delta virus and circular satellite RNAs depend for their multiplication on co-infection of the host by a helper virus; viroids replicate autonomously. Circular satellite RNAs and most known viroids are plant pathogens.

Hepatitis delta virus (HDV) is the only member of a "floating genus" of human hepatitis viruses. The virus is defective, needs the presence of hepatitis B virus (*Hepadnaviridae*) for replication, and borrows its envelope from this agent. Particles are spherical, 34 nm in diameter, and consist of an envelope of HBsAg and capsid-less helical nucleoprotein. Genomic RNA is 1.7 kb in size only. Genome replication occurs in the nucleus. During replication of genomic RNA, two species of complementary RNA are made, one of which is called the antigenome. The δ virus differs from viroids and plant satellite agents by the presence of δ antigen and dependence on hepatitis B virus.

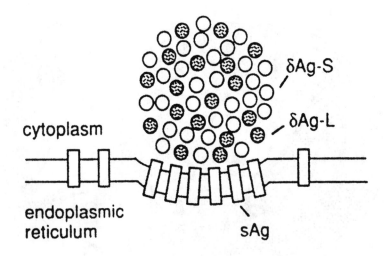

FIGURE 227
Model for HDV assembly. Antigens δAg-S and δAg-L (small- and large-envelope antigens of HDV, respectively) are shown as parts of a ribonucleoprotein complex which interacts with cytoplasmic domain(s) of hepadnavirus envelope proteins (HBsAg) after they have aggregated and been inserted into the membranes of the endoplasmic reticulum. The complex can make this interaction only if it contains δAg-L. (From Lazinski, D.W. and Taylor, J.M., *Adv. Virus Res.*, 43, 187, 1994. With permission.)

AGENTS WITH CIRCULAR ssRNA

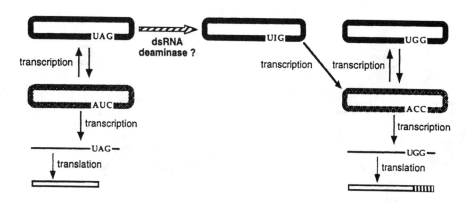

FIGURE 228

HDV replication. RNA editing plays a central role in the HDV life cycle. The target for editing is antigenomic RNA. The editing reaction is therefore a conversion of A to G and not of U to C. The editing is likely done by dsRNA adenosine deaminase. Thick black lines, HDV antigenomic RNA; cross-hatched lines, genomic RNA; thin lines, HDV antigenomic mRNA; open rectangle, first 195 amino acids of antigenomic RNA; vertically striped rectangle, additional 19 amino acids present at the C terminus of antigenomic p27 protein as a result of editing. (From Casey, J.L. and Gerin, J.L., *J. Virol.,* 69, 7593, 1995. With permission.)

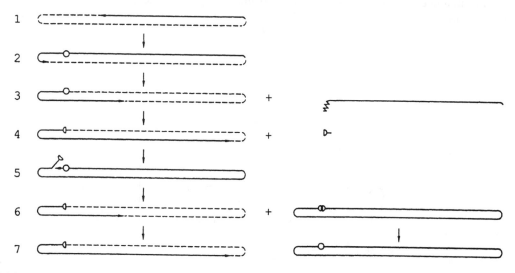

FIGURE 229

Abbreviated HDV genome replication strategy. Broken lines indicate the basic structure of the genomic RNA template. (1) Antigenomic RNA synthesis starts five nucleotides from one end of the rod. (2) The transcript proceeds beyond the self-cleavage site of the molecule (circle). (3) The transcript is processed like a cellular mRNA precursor and an mRNA is produced while transcription continues. (4) The continuing transcript self-cleaves, thereby stabilizing itself. (5) The transcript elongates further, copying once more the polyadenylation and self-cleavage sites. (6) Because of the ability of the RNA to fold the polyadenylation site into the RNA rod and because of the ability of δ antigen to bind to this structure, polyadenylation can be suppressed if there is enough δ antigen present. The only processing is thus self-cleavage and a full-length rod-shaped copy of the antigenome is released. (7) The antigenome circularizes via a self-ligation reaction. This RNA can then act as a template in a similar pathway to produce more circular genomic RNA. (Reprinted from Taylor, J.M., *Semin. Virol.,* 4, 313, 1993. By permission of the publisher Academic Press Ltd., London.)

AGENTS WITH CIRCULAR ssRNA

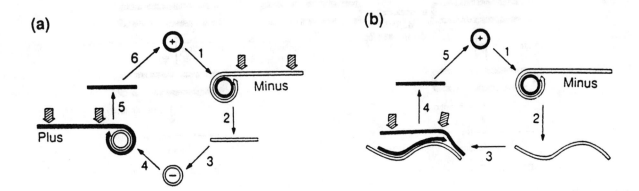

(a)

(b)

FIGURE 230

Two rolling-circle models for replication of circular pathogenic RNA. (a) Model where both (+) and (–) RNAs self-cleave. A continuous linear (–) strand produced by copying a circular (+) strand by an RNA polymerase is cleaved at specific sites (hatched arrows) to produce (–) monomers. These are circularized and copied to give a continuous linear (+) strand which is cleaved to linear monomers. The latter are circularized in turn to give circular progeny (+) strands. In the case of the satellite RNA of tobacco ringspot virus, only linear monomers are encapsidated. (b) Model where only the (+) RNA self-cleaves. The (+) strand is copied from a linear, oligomeric (–) strand. (A detailed version of model [a] may be found in Ref. 207; authors' note.) (Adapted from Symons, R.H., *Semin. Virol.,* 1, 117, 1990. By permission of the publisher Academic Press Ltd., London.)

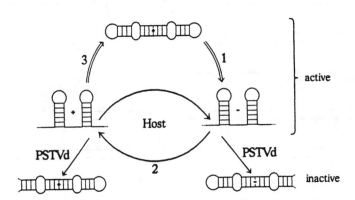

FIGURE 231

Replication cycle of potato spindle tuber viroid (PSTVd) showing critical secondary structures. (1) The inoculated rod-like and circular PSTVd is a template for synthesis of (–) strand oligomers. (2) During synthesis, (–) strands fold into a metastable multi-hairpin structure which is the template for synthesis of (+) strand oligomers and vice versa. (3) (+) strands undergo transitions into a processing structure (not shown) which is cleaved and ligated enzymatically to mature circles. Inactive side products are generated by rearrangement of multi-hairpin structures into thermodynamically favored rod-like linear oligomers. (From Gruner, R., Fels, A., Qu, F., Zimmat, R., Steger, G., and Riesner, D., *Virology,* 209, 60, 1995. With permission.)

8.C. PRIONS

No nucleic acid
No capsid, no envelope
Vertebrates

Prions are small proteinaceous infectious particles that resist procedures which degrade nucleic acids, but can be destroyed by enzymes or treatments which break down proteins. No nucleic acid and no capsid-like structures have been detected in association with prions; however, prions seem to be able to polymerize into rod-like structures. Prions cause spongiform encephalopathies in humans (kuru, Creutzfeldt–Jakob disease) and animals (scrapie of sheep, bovine spongiform encephalopathy, others). The present body of knowledge suggests that prions are composed in part or entirely of a protein designated PrP^{Sc} ("prion protein of scrapie") derived, perhaps by a change in conformation, from a normal cellular protein called PrP or PrP^{C}.

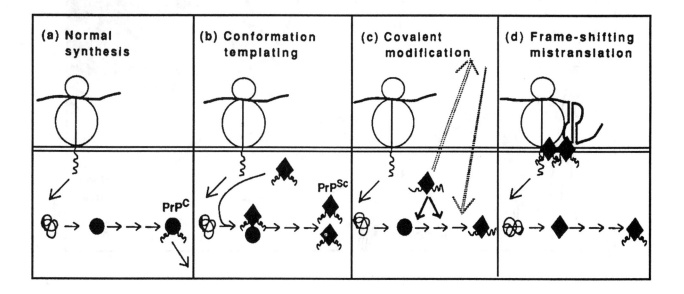

FIGURE 232

Models of prion replication. (a) The normal cellular form of prion protein (PrP^{C}) is synthetized by ribosomes on the membrane of the endoplasmic reticulum and extruded into the lumen where it is folded, covalently modified, and eventually transported to the plasma membrane. (b) Abnormal prion protein molecules (PrP^{Sc}) interact with PrP^{C} to form PrP^{C}-PrP^{Sc} dimers. PrP^{Sc} acts as a conformational template, converting PrP^{C} into novel Prp^{Sc}. (c) By acting directly and locally (solid arrows), perhaps by an allosteric effect, or indirectly and possibly remotely (dotted arrows), perhaps by expression of an otherwise silent gene, PrP^{Sc} interferes with the post-translational processing of newly synthesized PrP, bringing about the formation of new PrP^{Sc}. (d) Located in the membrane of the endoplasmic reticulum, PrP^{Sc} causes ribosomal frame-shifting during synthesis of PrP, by interacting with either nascent N-terminal PrP polypeptides or pseudoknots, in the mRNA encoding PrP. Normal post-translational processing produces new molecules of PrPS^{c}. (Reprinted from Wills, P.R., *Semin. Virol.,* 2, 181, 1991. By permission of the publisher Academic Press Ltd., London.)

PRIONS

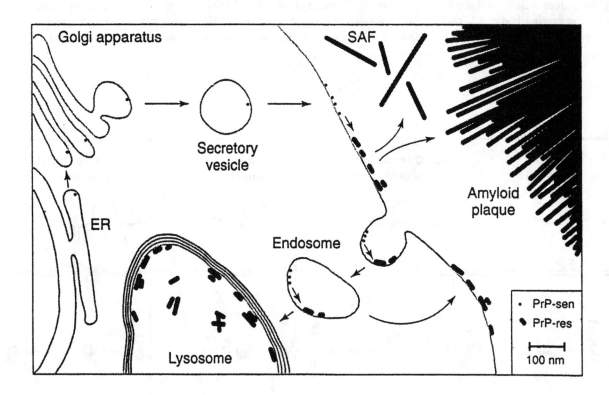

FIGURE 233

Generation of PrP-res (protease-resistant) from PrP-sen (protease-sensitive) on the plasma membrane and in endocytic vesicles. In some situations, smaller-sized PrP-res structures progress to generate scrapie-associated fibrils (SAF) and amyloid plaques. Scaled drawing illustrates how the smallest PrP-res structures might contain 20–100 PrP monomers and still be too small to visualize as fibrils by electron microscopy. ER, endoplasmic reticulum. (From Caughey, B.R. and Chesebro, B., *Trends Cell Biol.,* 7, 56, 1997 With permission.).

ABBREVIATIONS AND UNITS

Å	Ångström
ADP	Adenosine diphosphate
ATP	Adenosine triphosphate
cDNA	Complementary DNA
cRNA	Complementary RNA
DNA	Deoxyribonucleic acid
dNTP	Any deoxyribonucleotide precursor of DNA
dpi	Dots per inch
ds	Double-stranded
ER	Endoplasmic reticulum
gp	Gene product
h	Hour
IR	Inverted repeat
ITR	Inverted terminal repeat
K	Kilodalton
kb	Kilobase
kDa	Kilodalton
LTR	Long terminal repeat
min	Minute
mRNA	Messenger RNA
NAD	Nicotinamide adenine dinucleotide
nm	Nanometer
NP	Nucleoprotein
nt	Nucleotide
OH	Hydroxyl
ORF	Open reading frame
ori	Origin of replication
p, P	Protein
RER	Rough endoplasmic reticulum
RF	Replicative form
RFI, RFII	Replicative forms I or II
RI	Replicative intermediate
RNA	Ribonucleic acid
RNP	Ribonucleoprotein
S	Svedberg, sedimentation coefficient
ss	Single-stranded
ssb, SSB	Single-stranded DNA binding protein
tRNA	Transfer RNA
ts	Thermosensitive (in names of mutants, e.g., *ts*045)
VP	Viral protein
vRNA	Viral RNA
(+)	Plus, positive-sense, message-sense, messenger-sense
(−)	Minus, negative-sense, anti-messenger-sense

GLOSSARY

Abortive infection — Infection without production of infectious progeny
Adsorption — Attachment of a virus to a cell surface
Ambisense — Coding of proteins on both viral-sense and viral complementary RNAs
Antigenome — Complementary (+) RNA on which the negative strand is made
Antimessage — (–) stranded viral RNA which cannot act as mRNA and is transcribed to a (+) strand which acts as mRNA
Arbovirus — "Arthropod-borne," replicating in vertebrates and arthropods
Assembly — Assembly of virus components; also called morphogenesis or maturation
Attachment — Adsorption, fixation
Autophagolysosome — Vesicle derived from the endoplasmic reticulum and lysing a cellular organelle

Bacteriophage — Bacterial virus, synonym "phage"
Basophilic — Affinity for basic dyes
Bidirectional replication — Replication in two directions from a common origin
Budding — Virus release from a cellular membrane by formation of an outgrowth

C terminus — Carboxy terminal end of a polypeptide chain
Cap — Sequence of methylated bases at the 5′-terminal end of certain viral RNAs
Capsid — 1. Closed protein shell of virus particle
2. Head of tailed phages
Capsomer(e) — Morphological subunit of the capsid, visible in the electron microscope, formed of monomers (usually five or six)
Cell membrane — Plasma membrane
Cell wall — Rigid external covering of cytoplasmic membrane
Chaperonin — Nonstructural protein mediating the correct assembly of other proteins
Chemokine — Chemotactic cytokine that activates and directs leukocyte migration
Circular (cyclic) permutation — Variable location of DNA termini in a virus population
Circularization — End-to-end joining of linear DNA molecules
Cis-acting — Component of a DNA sequence which influences the activity of the same or an adjacent sequence
Cis-Golgi — Network between endoplasmic reticulum and Golgi stack
Cistron — Basic unit of genetic function; usually a gene
Clathrin — Protein forming a polyhedral lattice in pits and vesicles involved in endocytosis
Cleavage — Cutting of a nucleic acid or protein at specific sites
Coat — Protective layer(s) surrounding the viral nucleic acid
Cohesive ends — "Sticky ends"; complementary single-stranded ends of a dsDNA molecule
Competent — Ready to accept or act
Complementary — Single-stranded daughter strand synthetized on viral DNA or RNA
Concatemer (concatamer) — Giant molecule made up of multiple copies of covalently joined units
Conformation — Three-dimensional state
Conservative replication — Type of nucleic acid replication in which the parent double strand is preserved and both strands of the progeny molecule are newly synthetized
Constitutive — Permanently present, structural
Core — 1. Inner part of viral particle
2. Inner tube of contractile phage tails
Cos site — Cohesive end
Covalently — Two atoms linked by one or more pairs of electrons
Cowdry type A — Intranuclear inclusion body in cells infected with human herpesviruses
Creutzfeldt–Jakob disease — A transmissible human spongiform encephalopathy
Cubic symmetry — A regular polyhedron (tetrahedral, octahedral, icosahedral)

Cuticle	Thin exterior layer on outer surfaces of plants (main component cutin)
Cytoplasmic membrane	Bounding membrane of cytoplasm, plasma membrane
Cytoplasmic polyhedrosis	Insect disease characterized by large cytoplasmic polyhedral inclusion bodies
Cytosol	Cytoplasm without organelles and cytoskeleton (eukaryotes)
Dimer	Entity consisting of two units
Diploid	Provided with homologous paired chromosomes
Direct repeat	Repeat sequences in the same orientation
Dodecamer	Body consisting of 12 pieces
Domain	Segment identified by a structure or function
Eclipse	Time between disappearance of intracellular infecting viruses, or entry of naked nucleic acid, and appearance of intracellular progeny virions
Egestion	Exit of virus by budding or excretion
Egress	Same as above
Encapsidation	Inclusion of viral DNA or RNA into a capsid
Endocytosis	Intake via membrane-bound vesicles
Endoplasmic reticulum	Intracellular system of channels and vesicles (cisternae) contiguous with nuclear membrane (eukaryotes)
Endosome	Organelle of the endocytic pathway, located between plasma membrane and lysosomes
Env	Envelope gene or protein
Envelope	Lipoprotein membrane surrounding a viral capsid or nucleoprotein
Eosinophilic	Stainable reddish by eosin
Episome	Genetic element able to replicate independently in the cytoplasm or as integrated part of a cellular genome
Ergastoplasm	Granular endoplasmic reticulum (obsolete)
Eukaryote	Organism or cell provided with a nucleus (minimal definition)
Eukaryotic virus	Virus of an eukaryote
Exocytosis	Release of contents of secretory vesicles to the exterior
F-specific	Specific for F (fertility factor-related) sex pili
Fc	Crystallizable immunoglobulin fragment
Feulgen-positive	Stained by a DNA-specific (fuchsin sulfite) stain
Flip-flop	Movement of a molecule from one leaflet to the other of a bilayer membrane
Gag	Group-specific-antigen gene or protein
Genomic	Relating to the virus genome
Golgi apparatus	Stack of flattened cisternae forming an intracellular transport system in eukaryotes
Granulosis	Insect disease characterized by granular inclusion bodies
Guarnieri body	Cytoplasmic inclusion body in cells infected with vaccinia and related viruses
Gyrase	Enzyme introducing negative supercoils into a DNA helix
Hairpin loop	Double-stranded segment formed by base pairing of two regions of the same ssDNA or ssRNA
Helical symmetry	Structure in which the constitutive subunits form a helix
Helicase	Enzyme unwinding dsDNA or dsRNA
Hemagglutinin	Protein causing agglutination of red blood cells
Hemocoel	Blood-containing body cavity of insects
Hexamer	Group of six identical subunits
Hexon	Same as above
Holoenzyme	Enzyme complete with all structural domains, cofactors, and subunits required for function
Hydrophobic	Poorly or not at all soluble in water
Hyphae	Vegetative-phase (branching) filamentous tubes of fungi or algae
Icosahedron	Platonic solid with 20 triangular faces, 30 edges, 12 apices, and twofold, threefold, and fivefold axes of symmetry
Idiotype	Characteristic set of antigenic determinants in a specific antigen

Inclusion body	Microscopically visible intracellular aggregate of viruses and/or protein
Induction	Activation of a latent virus infection
Integration	Insertion of viral DNA into the host genome
Interferon	Type of (antiviral) regulatory proteins produced by vertebrate cells
Inverted repeat	Repeat sequences in opposite directions
Kuru	A transmissible human spongiform encephalopathy
Latent period	Time between virus infection and appearance of progeny virus
Ligand	Substance able to bind
Ligase	Enzyme joining two polynucleotide chains
Lysis	Disruption of host cell
Lysogeny	Carriage of a prophage by a bacterium, literally "generation of lysis"
Lysosome	Intracytoplasmic membranous vesicle containing hydrolytic enzymes
Lysozyme	Enzyme hydrolyzing certain glycosidic linkages of peptidoglycan
Lytic	Causing lysis
Matrix	"Mother substance," e.g.
	1. Protein located between viral envelope and nucleocapsid (orthomyxo-, paramyxo-, rhabdoviruses)
	2. Major protein of inclusion bodies
	3. Ground substance of viroplasm
	4. Template
Maturation	Assembly; also chemical or enzymatic processing toward a final state
-mer	Suffix derived from Greek *meros*, part (e.g., dimer, trimer, pentamer)
Message-sense	RNA that can be used without transcription (as mRNA)
Messenger RNA (mRNA)	Same as above; RNA translatable into protein
Minichromosome	Viral DNA assembled into chromatin with host histones
Monocistronic	RNA transcript derived from a single gene and coding for a single protein
Monomer	An individual subunit
Muralytic	Lysing a bacterial cell wall
N terminus	Amino terminal end of a polypeptide chain
Naked	No envelope
Negative-sense	(−) sense; nucleic acid complementary to the (+) strand of viral mRNA
Negri body	Cytoplasmic inclusion body in cells infected with rabies virus
Nested set	Hierarchical arrangement of related items within a similar item
Neuraminidase	Enzyme hydrolyzing glycosidic linkages between neuraminic acid and other molecules
Nick	Single-stranded gap in dsDNA
Novel	Newly produced, progeny
Nuclear polyhedrosis	Arthropod disease characterized by large intranuclear polyhedral inclusion bodies
Nucleocapsid	Capsid + nucleic acid
Nucleoid	Electron-dense central region of certain viruses
Nucleoplasm	Content of nucleus excluding the nucleolus
Occluded	Included within a protein crystal
One-step growth	Single cycle of replication in a cell population infected synchronously
Open reading frame (ORF)	DNA or RNA sequence between an initiation (start) and a termination (stop) codon; corresponds generally to a gene
Overlapping gene	Part of genome readable in more than one reading frame to give two or more proteins
Pararetrovirus	Virus practicing transcription of DNA to RNA (e.g., hepadnaviruses)
-partite	Consisting of parts
Penetration	Entry of virus, nucleocapsid, or viral nucleic acid into cell
Pentamer	Group of five identical subunits
Penton	1. Pentamer
	2. Complex structure of the adenovirus capsid consisting of a pentamer base, a fiber, and a terminal knob

Peplomer	Glycoprotein projection of viral envelope, also called "spike"
Peptidoglycan	Family of glycans cross-linked by peptides found in bacterial cell walls
Perinuclear	Surrounding the nucleus
Periplasm	Space between bacterial or yeast cell wall and plasma membrane
Permissive	Able to support a productive virus infection
Phage	Bacteriophage
Phagolysosome	Vesicle for digestion of internalized particles, formed by fusion of phagosomes and lysosomes
Phagosome	Phagocytic vesicle engulfing a virus or other particle
Pilus	Any filamentous bacterial appendage other than flagella
Pinocytosis	Cellular uptake of fluids and solutes by vacuoles; fluid-phase endocytosis
Plasmalemma	Cytoplasmic membrane (of fungi or protozoa)
Plasma membrane	Cytoplasmic membrane
Plasmid	Self-replicating extrachromosomal replicon, generally circular dsDNA
Plasmodesma	Cytoplasmic channel between adjacent plant cells
Pol	Polymerase gene or protein
Poly(A)	Polymer of adenosine nucleotides
Polyadenylation	Addition of a poly(A) tail to the 3′ end of RNA
Polyhedron	Body with multiple faces
Polymerase	Enzyme catalyzing DNA or RNA synthesis
Polyprotein	Precursor protein to be cleaved into several identical or different proteins
Portal protein	Protein located at one vertex of a phage head and linked to DNA packaging
Positive sense	(+) sense 1. RNA strand which functions as mRNA 2. DNA strand with the same sequence as mRNA
Pregenome	RNA precursor of hepatitis B virus DNA genome
Primase	Enzyme synthetizing RNA primers
Primer	Small RNA fragment, or protein, that serves as a starting point for DNA synthesis
Primosome	Assembly including a primase and other subunits
Prion	"Proteinaceous infectious particle"
Procapsid	Immature capsid
Processing	Cleavage
Progeny	Daughter generation in virus replication
Prohead	Procapsid of tailed phages
Prokaryote	Organism without a nucleus (minimal definition); bacterium
Prokaryotic virus	Bacterial virus
Prophage	Latent phage genome
Protomer	Intermediate stage in picornavirus assembly
Proton pump	Cellular membrane protein able to transport protons across the membrane
Provirus	1. dsDNA copy of an ssRNA retroviral genome 2. Any latent (integrated) virus genome 3. Immature virus (obsolete)
Receptor	Virus-binding site on cell surface
Receptosome	Endosome (obsolete)
Repeat	Nucleotide sequence occurring more than once within the same molecule
Replicase	RNA-dependent RNA polymerase; more generally an enzyme involved in replication of viral genomes
Replication	1. Multiplication of a virus 2. Production of a DNA or RNA strand from the original
Replicative form	Double-stranded form of ssDNA or RNA during replication
Replicative intermediate	Complex formed by simultaneous synthesis of one or more RNA strands from a single complementary strand
Reverse transcriptase	RNA-dependent DNA polymerase for DNA synthesis on RNA templates
Rise period	Time required for reaching a constant virus titer

Rolling-circle mechanism	Type of DNA or RNA replication in which a sole replication fork moves indefinitely around a circular template
Rough endoplasmic reticulum	Endoplasmic reticulum with attached ribosomes
Satellite RNA	Capsid-less RNA (ss or ds, linear or circular) that depends on a helper virus for replication
Satellite virus	Defective virus depending on a helper
Scaffolding protein	Protein around which a capsid is built
Scrapie	Transmissible spongiform encephalopathy of sheep
Segmented genome	Genome composed of two or more molecules of nucleic acid
Semi-conservative replication	Type of dsDNA or RNA replication in which both strands are templates, generating progeny viruses with a parent strand and a new strand
Sex pilus	Pilus involved in bacterial mating
Shell	Sum of rigid (lipo)protein layers surrounding viral nucleic acid
Spike	Peplomer; also short fixation structure on phage tails
Splicing (RNA)	RNA processing with removal of introns and ligation of exons
Strand displacement	Type of dsDNA or RNA replication in which one strand only is replicated, the other strand being displaced from the template
Structural protein	Protein that is part of the virus structure
Subgenomic	mRNA segment smaller than genomic RNA; not needed in natural infection, sometimes encapsidated, contains genes which are unavailable for translation from genomic RNA
Subviral	1. Smaller than a regular virus particle
	2. Breakdown product from *in vitro* treatment of viruses
	3. Immature virus (infrequent and inappropriate use)
Supercoiled	State of dsDNA in which the double helix is further twisted
Superhelical	Supercoiled
Syncytium	Multinucleate giant cell resulting from cell fusion
Tegument	Fibrous protein layer between capsid and envelope of herpesviruses
Template	Model DNA or RNA molecule from which a complementary nucleic acid strand is synthetized
Terminal redundancy	Nucleotide duplication at either end of a linear genome
Terminal repeats	Identical nucleotide sequences at both ends of the viral genome
Terminase	Enzyme cutting phage genomes at specific sites
Topoisomerase	Enzyme catalyzing the interconversion of topological DNA isomers
Toroid	Torus-like
Trans-acting	Gene whose product influences gene expression far away
Transcriptase	RNA polymerase transcribing RNA from DNA or RNA
Transcription	mRNA synthesis from a DNA or RNA template
Transformation	1. Gene transfer and expression by intake of "naked" exogenous DNA
	2. Induction of a tumor-like lifestyle in a cell
Trans-Golgi	Network between Golgi stack and plasma membrane
Translation	Protein synthesis directed by mRNA
Translocation	Transfer, movement
Transmembrane protein	Protein spanning the width of a membrane
Triangulation number	Number of equilateral triangles contained on the face of an icosahedron
Tripartite	Consisting of three parts or segments
Tumor antigen	Cell surface antigen indicative of tumor cells
Uncoating	Removal of outer layers of viral particles and exposure of viral nucleic acids
Viral DNA or RNA	DNA or RNA carried by a virion
Virion	Complete infectious particles
Virogenic stroma	Amorphous, membrane-less inclusion body associated with virus production
Viroid	Infectious, capsid-less, "naked" circular ssRNA
Viropexis	Form of pinocytosis with engulfment of whole viruses by vacuoles
Viroplasm	1. Amorphous inclusion body associated with virus infection
	2. Undefined substance within a virus (e.g., poxvirus)
Vitronectin	Serum protein with a role in adherence of mammalian cells to each other or to substrates

REFERENCES

1. **Ackermann, H.-W. and Berthiaume, L.,** *Atlas of Virus Diagrams,* CRC Press, Boca Raton, FL, 1987.
2. **Hull, R., Brown, F., and Payne, C.,** *Virology, Directory & Dictionary of Animal, Bacterial and Plant Viruses,* Macmillan, London, 1989.
3. **Kendrew, J., Ed.-in-chief,** *The Encyclopedia of Molecular Biology,* Blackwell Science, Oxford, 1994.
4. **Matthews, R.E.F.,** A history of viral taxonomy, in *A Critical Appraisal of Viral Taxonomy,* Matthews, R.E.F., Ed., CRC Press, Boca Raton, FL, 1983, 1.
5. **Murphy, F.A., Fauquet, C.M., Bishop, D.H.L., Ghabrial, S.A., Jarvis, A.W., Martelli, G.P., Mayo, M.A., and Summers, M.D., Eds.,** *Virus Taxonomy, Sixth Report of the International Committee on Taxonomy of Viruses,* Springer, Vienna, *Arch. Virol.,* Suppl. 10, 1995.
6. **Lwoff, A., Horne, R.W., and Turnier, P.,** A system of viruses, *Cold Spring Harbor Symp. Quant. Biol.,* 27, 51, 1962.
7. **Van Regenmortel, M.H.V.,** Virus species, a much overlooked but essential concept in virus classification, *Intervirology,* 31, 241, 1990.
8. **Ellis, E.L. and Delbrück, M.,** The growth of bacteriophage, *J. Gen. Physiol.,* 22, 365, 1939.
9. **Hershey, A.D. and Chase, M.,** Independent functions of viral protein and nucleic acid in growth of bacteriophage, *J. Gen. Physiol.,* 36, 39, 1952.
10. **Verduin, B.J.M.,** Early interactions between viruses and plants, *Semin. Virol.,* 3, 423, 1992.
11. **Baltimore, D.,** Expression of animal virus genomes, *Bacteriol. Rev.,* 35, 235, 1971.
12. **Lwoff, A.,** Lysogeny, *Bacteriol. Rev.,* 17, 269, 1953.
13. **Fenner, F., McAuslan, B.R., Mims, C.A., Sambrook, J., and White, D.O.,** *The Biology of Animal Viruses,* 2nd ed., Academic Press, New York, 1974, 177 and 342.
14. **Fenner, F.O. and White, D.O.,** *Medical Virology,* 2nd ed., Academic Press, New York, 1976, 50.
15. **Volk, W.A. and Wheeler, M.F.,** *Basic Microbiology,* 4th ed., J.B. Lippincott, Philadelphia, 1980, 144.
16. **Cann, A.J.,** Replication, in *Principles of Molecular Virology,* Cann, A.J., Ed., Academic Press, London, 1993, 86.
17. **Dulbecco, R. and Ginsberg, H.S.,** Adenoviruses, in *Microbiology*, 3rd ed., Davis, B.D., Dulbecco, R., Eisen, H.N., and Ginsberg, H.S., Eds., Harper & Row, Hagerstown, MD, 1980, 1047.
18. **Dales, S.,** Early events in cell–animal interactions, *Bacteriol. Rev.,* 37, 103, 1973.
19. **Pelczar, M.J., Chan, E.C.S., and Krieg, N.R.,** *Microbiology — Concepts and Applications,* McGraw-Hill, New York, 1993, 427 and 428.
20. **Lentz, T.L.,** The recognition event between virus and host cell receptor, *J. Gen. Virol.,* 71, 751, 1990.
21. **Dimmock, N.J. and Primrose, S.B.,** *Introduction to Modern Virology,* 3rd ed., Blackwell Scientific Publications, Oxford, 1987, 169.
22. **Pettersson, R.F.,** Protein localization and virus assembly at intracellular membranes, in *Protein Traffic in Eukaryotic Cells,* Compans, R.W., Ed., Springer, Berlin, *Curr. Topics Microbiol. Immunol.,* 170, 67, 1991.
23. **Dulbecco, R. and Ginsberg, H.S.,** Multiplication and genetics of animal viruses, in *Microbiology,* 3rd ed., Davis, B.D., Dulbecco, R., Eisen, H.N., and Ginsberg, H.S., Eds., Harper & Row, Hagerstown, MD, 1980, 967.
24. **Girard, M. and Hirth, L.,** *Virologie Générale et Moléculaire,* 2nd ed., Doin, Paris, 1989, 362, 365, and 369.
25. **Ackermann, H.-W.,** Les virus des bactéries, in *Virologie Médicale,* Maurin, J., Ed., Flammarion, Paris, 1985, 196.
26. **Fenner, F., Bachmann, P.A., Gibbs, E.P.J., Murphy, F.A., Studdert, M.J., and White, D.O.,** *Veterinary Virology,* Academic Press, Orlando, 1987, 64.
27. **Matthews, R.E.F.,** *Fundamentals of Plant Virology,* Academic Press, San Diego, 1992, 106 and 164.
28. **Samuel, C.E.,** Interferon, in *Encyclopedia of Microbiology,* Vol. 2, Lederberg, S., Ed.-in-chief, Academic Press, San Diego, 1992, 533.
29. **Marvin, D.A. and Wachtel, E.J.,** Structure and assembly of filamentous bacterial viruses, *Nature,* 253, 19, 1975.
30. **Webster, J.E. and Lopez, J.,** Structure and assembly of the class I filamentous bacteriophage, in *Virus Structure and Assembly,* Casjens, S., Ed., Jones and Bartlett, Boston, 1985, 235.
31. **Dunker, A.K., Ensign, L.D., Arnold, G.E., and Roberts, L.M.,** A model for fd phage penetration and assembly, *FEBS Lett.,* 292, 271, 1991.
32. **Model, P. and Russel, M.,** Filamentous bacteriophage, in *The Bacteriophages,* Vol. 2, Calendar, R., Ed., Plenum Press, New York, 1988, 375.

33. **Keppel, F., Fayet, O., and Georgopoulos, C.,** Strategies of bacteriophage DNA replication, in *The Bacteriophages,* Vol. 2, Calendar, R., Ed., Plenum Press, New York, 1988, 145.

34. **Horne, R.W.,** *The Structure and Function of Viruses,* Edward Arnold, London, 1978, 31, 35, and 50.

35. **Maniloff, J., Das, J., and Christensen, J.R.,** Viruses of mycoplasmas and spiroplasmas, *Adv. Virus Res.,* 21, 343, 1977.

36. **Casjens, S.,** Nucleic acid packaging by viruses, in *Virus Structure and Assembly,* Casjens, S., Ed., Jones and Bartlett, Boston, 1985, 75.

37. **Freifelder, D.,** *Molecular Biology, A Comprehensive Introduction to Prokaryotes and Eukaryotes,* Science Books International, Boston, 1983, 612, 616, 626, 653, 654, 659, 660, 662, and 672.

38. **Arai, K.-L., Low, R., Kobori, J., Shlomai, J., and Kornberg, A.,** Mechanism of *dnaB* protein action, *J. Biol. Chem.,* 256, 5273, 1981.

39. **Hayashi, M., Aoyama, A., Richardson, D.E., and Hayashi, M.N.,** Biology of the bacteriophage φX174, in *The Bacteriophages,* Vol. 2, Calendar, R., Ed., Plenum Press, New York, 1988, 1.

40. **Arai, N., Polder, L., Arai, K.-L., and Kornberg, A.,** Replication of φX174 DNA with purified enzymes. II. Multiplication of the duplex form by coupling of continuous and discontinuous synthetic pathways, *J. Biol. Chem.,* 256, 5239, 1981.

41. **Toolan, H.W. and Ellem, K.O.A.,** Parvoviridae, in *Virology and Rickettsiology,* Vol. I, Part 2, Hsiung, G.-D. and Green, R.H., Eds., CRC Press, Boca Raton, FL, 1978, 3.

42. **Berns, K.I., Bergoin, M., Bloom, M., Lederman, M., Muzyczka, N., Siegl, G., Tal, J., and Tattersall, P.,** Parvoviridae, in *Virus Taxonomy, Sixth Report of the International Committee on Taxonomy of Viruses,* Murphy, F.A., Fauquet, C.M., Bishop, D.H.L., Ghabrial, S.A., Jarvis, A.P., Martelli, G.P., Mayo, M.A., and Summers, M.D., Eds., Springer, Vienna, *Arch. Virol.,* Suppl. 10, 169, 1995.

43. **Bernhard, W.,** Fine structural lesions induced by viruses, in *Ciba Foundation Symposium on Cellular Injury,* De Reuck, A.V.S. and Knight, J., Eds., J. & A. Churchill, London, 1964, 209.

44. **Jawetz, E., Melnick, J.L., and Adelberg, E.A.,** *Review of Medical Microbiology,* 15th ed., Lange Medical Publications, Palo Alto, CA, 1982, 336.

45. **Dales, S.,** An electron microscope study of the early association between two mammalian viruses and their hosts, *J. Cell Biol.,* 13, 303, 1962.

46. **Chardonnet, Y. and Dales, S.,** Early event in the interaction of adenoviruses with HeLa cells. II. Comparative observations on the penetration of types 1, 5, 7, and 12, *Virology,* 40, 478, 1970.

47. **Nermut, M.V.,** Adenoviridae, in *Animal Virus Structure,* Nermut, M.V. and Steven, A.C., Eds., Elsevier, Amsterdam, 1987, 373.

48. **Lutz, P., Puvion-Dutilleul, F., Lutz, Y., and Kedinger, C.,** Nucleoplasmic and nucleolar distribution of the adenovirus IVa2 gene product, *J. Virol.,* 70, 3449, 1996.

49. **D'Halluin, J.C.,** Virus assembly, in *The Molecular Biology of Adenoviruses 1,* Doerfler, W., Ed., Springer, Berlin, *Curr. Topics Microbiol. Immunol.,* 109(1), 47, 1995.

50. **Friefeld, B.R., Lichy, J.H., Field, J., Gronostajski, R.M., Guggenheimer, R.A., Krevolin, M.D., Nagata, K., Hurwitz, J., and Horwitz, M.S.,** The *in vitro* replication of adenovirus DNA, in *The Molecular Biology of Adenoviruses 2,* Doerfler, W., Ed., Springer, Berlin, *Curr. Topics Microbiol. Immunol.,* 110(2), 221, 1984.

51. **Brookes, S.M., Dixon, L.K., and Parkhouse, R.M.E.,** Assembly of African swine fever virus: quantitative ultrastructural analysis *in vitro* and *in vivo,* *Virology,* 224, 84, 1996.

52. **Volkman, L.E.,** The 64K envelope protein of budded *Autographa californica* nuclear polyhedrosis virus, in *The Molecular Biology of Baculoviruses,* Doerfler, W. and Böhm, P., Eds., Springer, Berlin, *Curr. Topics Microbiol. Immunol.,* 131, 103, 1986.

53. **Blissard, G.W. and Rohrmann, G.F.,** Baculovirus diversity and molecular biology, *Annu. Rev. Entomol.,* 35, 127, 1990.

54. **Arnott, H.J. and Smith, K.M.,** An ultrastructural study of the development of a granulosis virus in the cells of the moth *Plodia interpunctella* (Hbn.), *J. Ultrastruct. Res.,* 21, 251, 1967.

55. **Funk, C.J. and Consigli, R.A.,** Phosphate cycling on the basic protein of *Plodia interpunctella* granulosis virus, *Virology,* 193, 396, 1993.

56. **Roizman, B. and Sears, A.E.,** Herpes simplex viruses and their replication, in *Virology,* 2nd ed., Fields, B.N. and Knipe, D.M., Eds.-in-chief, Raven Press, New York, 1990, 1795.

57. **Jahn, G. and Plachter, B.,** Diagnostics of persistent viruses: human cytomegalovirus as an example, *Intervirology,* 35, 60, 1993.

58. **De Thé, G. and Lenoir, G.,** Comparative diagnosis of Epstein-Barr virus-related diseases: infectious mononucleosis, Burkitt's lymphoma, and nasopharyngeal carcinoma, in *Comparative Diagnosis of Viral Diseases,* Vol. I, Part A, Kurstak, E. and Kurstak, C., Eds., Academic Press, New York, 1977, 195.

59. **Topilko, A. and Michelson, S.,** Hyperimmediate entry of human cytomegalovirions and dense bodies into human fibroblasts, *Res. Virol.,* 145, 75, 1994.

60. **Gershon, A.A., Sherman, D.L., Zhu, Z., Gabel, C.A., Ambron, M.T., and Gershon, M.,** Intracellular transport of newly synthetized varicella-zoster virus: final envelopment in the *trans*-Golgi network, *J. Virol.,* 68, 6372, 1994.

61. **Roffman, E., Albert, J.P., Goff, J.P., and Frenkel, N.,** Putative site for the acquisition of human herpesvirus 6 virion tegument, *J. Virol.,* 64, 6308, 1990.

62. **Thomsen, D.R., Newcomb, W.W., Brown, J.C., and Homa, F.L.,** Assembly of the herpes simplex virus capsid: requirement for the carboxyl-terminal twenty-five amino acids of the proteins encoded by the UL26 and the UL26.5 genes, *J. Virol.,* 69, 3690, 1995.

63. **Ben-Porat, T.,** Replication of herpesvirus DNA, *Curr. Topics Microbiol. Immunol.,* 91, 81, 1981.

64. **Severini, A., Scraba, D.G., and Tyrrell, D.L.J.,** Branched structures in the intracellular DNA of herpes simplex virus type 1, *J. Virol.,* 70, 3169, 1996.

65. **Hammerschmidt, W. and Mankertz, J.,** Herpesviral replication: between the known and the unknown, *Semin. Virol.,* 2, 257, 1991.

66. **Murti, K.G., Goorha, R., and Granoff, A.,** An usual replication strategy of an animal iridovirus, *Adv. Virus Res.,* 30, 1, 1985.

67. **Oxman, M.N.,** Papovaviridae, in *Virology and Rickettsiology,* Vol. I, Part 2, Hsiung, G.-D. and Green, R.H., Eds., CRC Press, Boca Raton, FL, 1978, 17.

68. **Consigli, R.A. and Center, M.S.,** Recent advances in polyoma virus research, *CRC Crit. Rev. Microbiol.,* 6, 263, 1978.

69. **Milavetz, B. and Hopkins, T.,** Simian virus 40 encapsidation: characteristics of early intermediates, *J. Virol.,* 43, 830, 1982.

70. **DePamphilis, M.L., Anderson, S., Bar-Shavit, R., Collins, E., Edenberg, H., Herman, T., Karas, B., Kaufman, G., Krokan, H., Shelton, E., Su, R., Taper, D., and Wassarman, P.M.,** Replication and structure of simian virus 40 chromosomes, *Cold Spring Harbor Symp. Quant. Biol.,* 43, 679, 1978.

71. **Moss, B.,** Poxviridae and their replication, in *Virology,* 2nd ed., Vol. 2, Fields, B.N. and Knipe, D.M., Eds.-in-chief, Raven Press, New York, 1990, 2079.

72. **Dales, S.,** The uptake and development of vaccinia virus in strain L cells followed with labelled viral deoxyribonucleic acid, *J. Cell Biol.,* 18, 51, 1963.

73. **Müller, G. and Williamson, J.D.,** Poxviridae, in *Animal Virus Structure,* Nermut, M.V. and Steven, A.C., Eds., Elsevier, Amsterdam, 1987, 421.

74. **Palmer, E.L. and Martin, M.L.,** *Electron Microscopy in Viral Diagnosis,* CRC Press, Boca Raton, FL, 1988, 65, 99, 108, 114, 139, and 175.

75. **VanSlyke, J.K. and Hruby, D.E.,** Posttranslational modification of vaccinia virus proteins, in *Poxviruses,* Moyer, R.W. and Turner, P.C., Eds., Springer, Berlin, *Curr. Topics Microbiol. Immunol.,* 163, 185, 1990.

76. **Scherrer, R.,** Morphogénèse et ultrastructure du virus fibromateux de Shope, *Pathol. Microbiol.,* 31, 129, 1968.

77. **Devauchelle, G., Bergoin, M., and Vago, C.,** Etude ultrastructurale du cycle de réplication d'un entomopoxvirus dans les hémocytes de son hôte, *J. Ultrastruct. Res.,* 37, 301, 1971.

78. **Traktman, P.,** Molecular genetic and biochemical analysis of poxvirus DNA replication, *Semin. Virol.,* 2, 291, 1991.

79. **Tuma, R., Bamford, J.K.H., Bamford, D.H., Russell, M.P., and Thomas, G.J.,** Structure, interactions and dynamics of PRD1 virus. I. Coupling of subunit foldings and capsid assembly, *J. Mol. Biol.,* 257, 87, 1996.

80. **Bamford, D.H., Caldentey, J., and Bamford, J.K.H.,** Bacteriophage PRD1: a broad host range dsDNA tectivirus with an internal membrane, *Adv. Virus Res.,* 45, 281, 1995.

81. **Ackermann, H.W.,** Frequency of morphological phage descriptions in 1995, *Arch. Virol.,* 141, 209, 1996.

82. **Evans, E.A.,** Bacteriophage as nucleoprotein, *Fed. Proc.,* 15, 827, 1956.

83. **Penso, G.,** Attaque et démolition de la cellule bactérienne par les phages, in *Problèmes Actuels de Virologie,* Hauduroy, P., Ed., Masson, Paris, 1954, 107.

84. **Wood, W.B. and Edgar, R.S.,** Building a bacterial virus, *Sci. Amer.,* 217, 61, 1967.

85. **Jacob, F.,** *Les Bactéries Lysogènes et la Notion de Provirus,* Masson, Paris, 1954, 30.

86. **Hercík, F.,** *Biophysik der Bakteriophagen,* VEB Deutscher Verlag der Wissenschaften, Berlin, 1959, 5.

87. **Bertani, G.,** Lysogeny, *Adv. Virus Res.,* 5, 151, 1958.

88. **Campbell, A.,** Episomes, *Adv. Genet.,* 11, 101, 1962.

89. **Hendrix, R.W.,** Shape determination in virus assembly: the bacteriophage example, in *Virus Structure and Assembly,* Casjens, S., Ed., Jones and Bartlett, Boston, 1985, 169.

90. **Black, L.W.,** DNA packaging and cutting by phage terminases: control in phage T4 by a synaptic mechanism, *BioEssays,* 17, 1025, 1995.

91. **Hud, N.V.,** Double-stranded DNA organization in bacteriophage heads: an alternative toroid-based model, *Biophys. J.,* 69, 1355, 1995.

92. **Boyd, R.F.,** *General Microbiology,* Times Mirror/Mosby College, St. Louis, 1984, 358.

93. **Mathews, C.K.,** An overview of the T4 development program, in *Molecular Biology of Bacteriophage T4,* Karam, J.D., Ed.-in-chief, American Society for Microbiology, Washington, D.C., 1994, 1.

94. **Kutter, E.,** T4 synthesis as related to the T4 genetic map, in *Molecular Biology of Bacteriophage T4,* Karam, J.D., Ed.-in-chief, American Society for Microbiology, Washington, D.C., 1994, ii.

95. **Furukawa, H., Kuroiwa, T., and Mizushima, S.,** DNA injection during bacteriophage T4 infection of *Escherichia coli, J. Bacteriol.,* 154, 938, 1983.

96. **Dreiseikelmann, B.,** Translocation of DNA across bacterial membranes, *Microbiol. Rev.,* 58, 293, 1994.

97. **Wood, W.B., Dickson, R.C., Bishop, R.J., and Revel, H.R.,** Self-assembly and non-self-assembly in bacteriophage T4 morphogenesis, in *Generation of Subcellular Structure,* First John Innes Symposium, Norwich, 1973, Markham, R., Ed., Elsevier, Amsterdam, 1973, 25.

98. **Dulbecco, R. and Ginsberg, H.S.,** Multiplication and genetics of bacteriophages, in *Microbiology,* 3rd ed., Davis, B.D., Dulbecco, R., Eisen, H., and Ginsberg, H.S., Eds., Harper & Row, Hagerstown, MD, 1980, 885.

99. **Black, L.M., Showe, M.K., and Steven, A.C.,** Morphogenesis of the T4 head, in *Molecular Biology of Bacteriophage T4,* Karam, J.D., Ed.-in-chief, American Society for Microbiology, Washington, D.C., 1994, 218.

100. **Hellen, C.U.T. and Wimmer, E.,** The role of proteolytic processing in the morphogenesis of virus particles, *Experientia,* 48, 201, 1992.

101. **Aebi, U., Bijlenga, R., Van der Broek, J., Eiserling, F., Kellenberger, C., Kellenberger, E., Mesyanzhinov, V., Müller, L., Showe, M., Smith, R., and Steven, A.,** The transformation of τ particles into T4 heads. II. Transformations of the surface lattice and related observations on form determination, *J. Supramol. Struct.,* 2, 253, 1974.

102. **Coombs, D.H. and Arisaka, F.,** T4 tail structure and function, in *Molecular Biology of Bacteriophage T4,* Karam, J.D., Ed.-in-chief, American Society for Microbiology, Washington, D.C., 1994, 259.

103. **Bishop, R.J. and Wood, W.B.,** Genetic analysis of T4 tail fiber assembly. I. A gene *37* mutation that allows bypass of gene *38* function, *Virology,* 72, 244, 1976.

104. **Dannenberg, R. and Mosig, G.,** Early intermediates in bacteriophage T4 replication and recombination, *J. Virol.,* 45, 813, 1983.

105. **Grimaud, R.,** Bacteriophage Mu head assembly, *Virology,* 217, 200, 1996.

106. **Cohen, G.,** Electron microscopy study of early lytic replication forms of bacteriophage P1 DNA, *Virology,* 131, 159, 1983.

107. **Dehò, G., Bertani, G., and Polissi, A.,** Bacteriophage P4-derived shuttle vectors for cloning and transposon mutagenesis in *Pseudomonas putida* and other Gram-negative bacteria, in *Pseudomonas: Molecular Biology and Biotechnology,* Galli, E., Silver, S., and Witholt, B., Eds., American Society for Microbiology, Washington, D.C., 1992, 358.

108. **Marvik, O.J., Dokland, T., Nøkling, R.H., Jacobsen, E., Larsen, T., and Lindqvist, B.H.,** The capsid size-determining protein Sid forms an external scaffold on phage P4 procapsids, *J. Mol. Biol.,* 259, 59, 1995.

109. **Furth, M.F. and Wickner, S.H.,** Lambda DNA replication, in *Lambda II,* Hendrix, R.W., Roberts, J.W., Stahl, F.W., and Weisberg, R.A., Eds., Cold Spring Harbor Laboratory, Cold Spring Harbor, NY, 1983, 145.

110. **Hohn, T. and Katsura, I.,** Structure and assembly of bacteriophage lambda, *Curr. Topics Microbiol. Immunol.,* 78, 69, 1977.

111. **Georgopoulos, C., Tilly, K., and Casjens, S.,** Lambdoid phage head assembly, in *Lambda II,* Hendrix, R.W., Roberts, J.W., Stahl, F.W., and Weisberg, R.A., Eds., Cold Spring Harbor Laboratory, Cold Spring Harbor, NY, 1983, 279.

112. **Zachary, A. and Simon, L.D.,** Size changes among λ capsid precursors, *Virology,* 81, 107, 1977.

113. **Hendrix, R.W.,** Tail length determination in double-stranded DNA bacteriophages, in *The Molecular Biology of Bacterial Virus Systems,* Hobom, G. and Rott, R., Eds., Springer, Berlin, *Curr. Topics Microbiol. Immunol.,* 136, 21, 1988.

114. **Catalano, C.E., Cue, D., and Feiss, M.,** Virus DNA packaging: the strategy used by phage λ, *Mol. Microbiol.,* 16, 1075, 1995.

115. **Dulbecco, R. and Ginsberg, H.S.,** Lysogeny and transducing bacteriophages, in *Microbiology,* 3rd ed., Davis, B.D., Dulbecco, R., Eisen, H.N., and Ginsberg, H.S., Eds., Harper & Row, Hagerstown, MD, 1980, 919.

116. **Echols, H. and Murialdo, H.,** Genetic map of bacteriophage lambda, *Microbiol. Rev.,* 42, 577, 1978.

117. **Majumdar, S., Dey, S.N., Chowdhury, R., Dutta, C., and Das, J.,** Intracellular development of choleraphage Φ149 under permissive and nonpermissive conditions, *Intervirology,* 29, 27, 1988.

118. **Roeder, G.S. and Sadowski, P.D.,** Bacteriophage T7 morphogenesis: phage-related particles in cells infected with wild-type and mutant T7 phage, *Virology,* 76, 263, 1977.

119. **Son, M., Watson, R.H., and Serwer, P.,** The direction and rate of bacteriophage T7 DNA packaging *in vitro, Virology,* 196, 282, 1993.

120. **Hendrix, R.W. and Garcea, R.L.,** Capsid assembly of dsDNA viruses, *Semin. Virol.,* 5, 15, 1994.

121. **Casjens, S. and King, J.,** P22 morphogenesis. I. Catalytic scaffolding protein in capsid assembly, *J. Supramol. Struct.,* 2, 202, 1974.

122. **Prasad, B.V.V., Prevelige, P.E., Marietta, E., Chen, R.O., Thomas, D., King, J., and Chiu, W.,** Three-dimensional transformation of capsids associated with genome packaging in a bacterial virus, *J. Mol. Biol.,* 231, 65, 1993.

123. **Anderson, D. and Reilly, B.,** Morphogenesis of bacteriophage Φ29, in *Bacillus subtilis,* Sonenshein, A.L., Hoch, J.A., and Losick, R., Eds., American Society for Microbiology, Washington, D.C., 1993, 859.

124. **Lee, C.-S. and Guo, P.,** *In vitro* assembly of infectious virions of double-stranded phage ϕ29 from cloned gene products and synthetic nucleic acids, *J. Virol.,* 69, 5018, 1995.

125. **Lee, C.-S. and Guo, P.,** Sequential interactions of structural proteins in phage ϕ29 procapsid assembly, *J. Virol.,* 69, 5024, 1995.

126. **Inciarte, M.R., Salas, M., and Sogo, J.M.,** Structure of replicating DNA molecules of *Bacillus subtilis* bacteriophage ϕ29, *J. Virol.,* 34, 187, 1980.

127. **Salas, M., Freire, R., Soengas, M.S., Esteban, J.A., Méndez, J., Bravo, A., Serrano, M., Blasco, M.A., Lázaro, J.M., Blanco, L., Gutiérrez, C., and Hermoso, J.M.,** Protein nucleic acid interactions in bacteriophage ϕ29 DNA replication, *FEMS Microbiol. Rev.,* 17, 73, 1995.

128. **Hohn, T. and Fütterer, J.,** Pararetroviruses and retroviruses: a comparison of expression strategies, *Semin. Virol.,* 2, 55, 1991.

129. **Lockhart, B.E.L., Olszewski, N.E., and Hull, R.,** Genus *Badnavirus,* in *Virus Taxonomy, Sixth Report of the International Committee on Taxonomy of Viruses,* Murphy, F.A., Fauquet, C.M., Bishop, D.H.L., Ghabrial, S.A., Jarvis, A.W., Martelli, G.P., Mayo, M.A., and Summers, M.D., Eds., Springer, Vienna, *Arch. Virol.,* Suppl. 10, 185, 1995.

130. **Covey, S.N.,** Pathogenesis of a plant retrovirus: CaMV, *Semin. Virol.,* 2, 151, 1991.

131. **Hoofnagle, G.H. and Di Bisceglie, A.M.,** Antiviral therapy of viral hepatitis, in *Antiviral Agents and Diseases of Man,* 3rd ed., Galasso, G.J., Whitley, R.J., and Merigan, T.C., Eds., Raven Press, New York, 1990, 415.

132. **Summers, J. and Mason, W.S.,** Replication of the genome of a hepatitis B-like virus by reverse transcription of an RNA intermediate, *Cell,* p. 403, 1982.

133. **Will, H., Reiser, W., Pfaff, E., Büscher, M., Sprengel, R., Cattaneo, R., and Schaller, H.,** Replication strategy of human hepatitis B virus, *J. Virol.,* 61, 904, 1987.

134. **Nassal, M. and Rieger, A.,** A bulged region of the hepatitis B virus RNA encapsidation signal contains the replication origin for discontinuous first-strand DNA synthesis, *J. Virol.,* 70, 2764, 1996.

135. **Gelderblom, H.R., Gentile, M., Scheidler, A., Özel, M., and Pauli, G.,** Zur Struktur und Funktion bei HIV: Gesichertes, neue Felder und offene Fragen, *AIDS Forsch.,* 8, 231, 1993.

136. **Gelderblom, H.R., Özel, M., and Pauli, G.,** Morphogenesis and morphology of HIV, *Arch. Virol.,* 106, 1, 1989.

137. **Arnold, E. and Ferstandig Arnold, G.,** Human immunodeficiency virus structure: implications for antiviral design, *Adv. Virus Res.,* 39, 1, 1991.

138. **Coffin, J.M.,** Retroviridae and their replication, in *Virology,* 2nd ed., Vol. 2, Fields, B.N. and Knipe, D.M., Eds.-in-chief, Raven Press, New York, 1990, 1437.

139. **Boothe, A.D., Van der Maaten, M.J., and Malmquist, W.A.,** Morphological variation of a syncytial virus from lymphosarcomatous and apparently normal cattle, *Arch. Ges. Virusforsch.,* 31, 373, 1970.

140. **Greene, W.C.,** The molecular biology of human immunodeficiency virus type 1 infection, *New England J. Med.,* 324, 308, 1991.

141. **Arts, E.J. and Wainberg, M.A.,** Human immunodeficiency virus type 1 reverse transcriptase and early events in transcription, *Adv. Virus Res.,* 46, 97, 1996.

142. **Brown, P.O.,** Integration of retroviral DNA, in *Retroviruses,* Swanstrom, K. and Vogt, P.K., Eds., Springer, Berlin, *Curr. Topics Microbiol. Immunol.,* 157, 19, 1990.

143. **Geleziunas, R., Bour, S., and Wainberg, M.A.,** Human immunodeficiency virus type 1-associated CD4 downmodulation, *Adv. Virus Res.,* 44, 203, 1994.

144. **Greene, W.C. and Cullen, B.R.,** The Rev-Rex connection: convergent strategies for the post-transcriptional regulation of HIV-1 and HTLV-1 gene expression, *Semin. Virol.,* 1, 195, 1990.

145. **Wong-Staal, F.,** Human immunodeficiency viruses and their replication, in *Virology,* 2nd ed., Vol. 2, Fields, B.N. and Knipe, D.M., Eds.-in-chief, Raven Press, New York, 1990, 1529.

146. **Palukaitis, P., Roossinck, M.J., Dietzgen, R.G., and Francki, R.I.B.,** Cucumber mosaic virus, *Adv. Virus Res.,* 41, 281, 1992.

147. **Sturman, L.S. and Holmes, K.V.,** The molecular biology of coronaviruses, *Adv. Virus Res.,* 28, 35, 1983.

148. **Holmes, K.V.,** Coronaviridae and their replication, in *Virology,* 2nd ed., Vol. 1, Fields, B.N. and Knipe, D.M., Eds.-in-chief, Raven Press, New York, 1990, 841.

149. **Fenger, T.W.,** Replication of RNA viruses, in *Textbook of Human Virology,* 2nd ed., Belshe, R.B., Ed., Mosby-Year Book, St. Louis, 1991, 74.

150. **Hase, T., Summers, P.L., and Houston Cohen, W.,** A comparative study of entry modes into C6/36 cells by Semliki Forest and Japanese encephalitis viruses, *Arch. Virol.,* 108, 101, 1989.

151. **Hase, T., Summers, P.L., Eckels, K.H., and Baze, W.B.,** Maturation process of Japanese encephalitis virus in cultured mosquito cells *in vitro* and mouse brain cells *in vivo, Arch. Virol.,* 96, 135, 1987.

152. **Beumer, J., Hannecart-Pokorni, E., and Godard, C.,** Bacteriophage receptors, *Bull. Inst. Pasteur,* 82, 173, 1984.

153. **Gallagher, T.M. and Rueckert, R.R.,** Assembly-dependent maturation cleavage in provirions of a small icosahedral insect ribovirus, *J. Virol.,* 62, 3399, 1988.

154. **Schneemann, A., Gallagher, T.M., and Rueckert, R.R.,** Reconstitution of flock house provirions: a model system for studying structure and assembly, *J. Virol.,* 68, 4547, 1994.

155. **Melnick, J.L.,** Picornaviruses: enteroviruses — polioviruses, in *Virology and Rickettsiology,* Vol. I, Part 1, Hsiung, G.-D. and Green, R.H., Eds., CRC Press, Boca Raton, FL, 1978, 111.

156. **Rueckert, R.R.,** Picornaviridae and their replication, in *Virology,* 2nd ed., Vol. 1, Fields, B.N. and Knipe, D.M., Eds.-in-chief, Raven Press, New York, 1990, 507.

157. **Crowell, R.L.,** Cellular receptors in virus infection, *ASM News,* 53, 422, 1987.

158. **Crowell, R.L. and Siak, J.-S.,** Receptor for group B coxsackieviruses: characterization and extraction from HeLa cell membranes, in *Perspectives in Virology,* Vol. 19, Pollard, M., Ed., Raven Press, New York, 1978, 39.

159. **Bienz, K., Bienz-Isler, G., Egger, D., Weiss, M., and Loeffler, H.,** Coxsackievirus infection in skeletal muscles of mice. II. Appearance and fate of virus progeny, *Arch. Ges. Virusforsch.,* 31, 251, 1970.

160. **Ansardi, D.C., Porter, D.C., Anderson, M.J., and Morrow, C.,** Poliovirus assembly and encapsidation of genomic RNA, *Adv. Virus Res.,* 46, 1, 1996.

161. **Joklik, W.K.,** The virus multiplication cycle, in *Zinsser Microbiology,* 17th ed., Joklik, W.K., Willett, H.P., and Amos, D.B., Eds., Appleton-Century-Crofts, New York, 1980, 1040.

162. **Butler, P.J.G., Bloomer, A.C., Bricogne, G., Champness, J.N., Graham, J., Guilley, H., Klug, A., and Zimmern, D.,** Tobacco mosaic virus assembly — specificity and the transition in protein structure during RNA packaging, in *Structure–Function Relationship of Proteins, Proc. Third John Innes Symposium,* Markham, R. and Horne, R.W., Eds., North-Holland, Amsterdam, 1976, 101.

163. **Butler, P.J.G.,** The current picture of the structure and assembly of tobacco mosaic virus, *J. Gen. Virol.,* 65, 253, 1984.

164. **Bykovsky, H.F., Yershov, F.I., and Zhdanov, V.M.,** Morphogenesis of Venezuelan equine encephalomyelitis virus, *J. Virol.,* 4, 496, 1969.

165. **Kielian, M.,** Membrane fusion and the alphavirus life cycle, *Adv. Virus Res.,* 45, 113, 1995.

166. **Mauracher, C.A., Gillam, S., Shukin, R., and Tingle, A.J.,** pH-dependent solubility shift of rubella virus capsid protein, *Virology,* 181, 773, 1991.

167. **Marsh, M. and Helenius, A.,** Virus entry into animal cells, *Adv. Virus Res.,* 36, 107, 1989.

168. **Simons, K. and Garoff, H.,** The budding mechanism of enveloped animal viruses, *J. Gen. Virol.,* 50, 1, 1980.

169. **Simons, K., Garoff, H., and Helenius, A.,** The glycoproteins of the Semliki Forest virus membrane, in *Membrane Proteins and Their Interactions with Lipids,* Vol. 1, Capaldi, R., Ed., Marcel Dekker, New York, 1977, 207.

170. **Bredenbeek, P.J. and Rice, C.M.,** Animal RNA virus expression systems, *Semin. Virol.,* 3, 297, 1992.

171. **Matthews, R.E.F.,** *Plant Virology,* 3rd ed., Academic Press, San Diego, 1991, 238.

172. **Salvato, M.S.,** Molecular biology of the prototype arenavirus, lymphocytic choriomeningitis virus, in *The Arenaviridae,* Salvato, M.S., Ed., Plenum Press, New York, 1993, 133.

173. **Schmaljohn, C.S. and Patterson, J.L.,** Bunyaviridae and their replication. II. Replication of Bunyaviridae, in *Virology,* 2nd ed., Vol. 1, Fields, B.N. and Knipe, D.M., Eds.-in-chief, Raven Press, New York, 1990, 1175.

174. **Matsuoka, Y., Chen, S.Y., and Compans, R.W.,** Bunyavirus protein transport and assembly, in *Bunyaviridae,* Kolakovsky, D., Ed., Springer, Berlin, *Curr. Topics Microbiol. Immunol.,* 169, 161, 1991.

175. **Goldbach, R. and De Haan, P.,** Prospects of engineered forms of resistance against tomato spotted wilt virus, *Semin. Virol.,* 4, 381, 1993.

176. **Bishop, D.H.L.,** The genetic basis for describing viruses as species, *Intervirology,* 24, 79, 1985.

177. **Palese, P. and Ritchey, M.B.,** Myxoviridae: orthomyxoviruses — influenza viruses, in *Virology and Rickettsiology,* Vol. I, Part 1, Hsiung, G.-D. and Green, R.H., Eds., CRC Press, Boca Raton, FL, 1978, 337.

178. **Huraux, J.M., Nicolas, J.C., and Agut, H.,** *Virologie,* Flammarion, Paris, 1985, 168.

179. **Outlaw, M.C. and Dimmock, N.J.,** Insights into neutralization of animal viruses gained from study of influenza virus, *Epidemiol. Infect.,* 106, 205, 1991.

180. **Compans, R.W., Meier-Ewert, H., and Palese, P.,** Assembly of lipid-containing viruses, *J. Supermol. Struct.,* 2, 496, 1974.

181. **Kilbourne, E.D.,** *Influenza,* Plenum Medical Books, New York, 1987, 127.

182. **Fenner, F.,** *The Biology of Animal Viruses,* Vol. I, Academic Press, New York, 1968, 264.

183. **Berthiaume, L., Joncas, J., and Pavilanis, V.,** Comparative structure, morphogenesis and biological characteristics of the respiratory syncytial (RS) virus and the pneumonia virus of mice (PVM), *Arch. Ges. Virusforsch.,* 45, 39, 1974.

184. **Georges, J.-C.,** Les paramyxoviridés: caractères généraux, in *Virologie Médicale,* Maurin, J., Ed., Flammarion, Paris, 1985, 475.

185. **Wagner, R.R.,** Rhabdovirus biology and infection — an overview, in *The Rhabdoviridae,* Wagner, R.R., Ed., Plenum Press, New York, 1987, 9.

186. **Danglot, C.,** La multiplication des virus à ARN, in *Virologie Médicale,* Maurin, J., Ed., Flammarion, Paris, 1985, 57.

187. **Knipe, D.M., Baltimore, D., and Lodish, H.F.,** Maturation of viral proteins in cells infected with temperature-sensitive mutants of vesicular stomatitis virus, *J. Virol.,* 21, 1149, 1977.

188. **Gosztonyi, G.,** Reproduction of lyssavirus: ultrastructural composition of lyssavirus and functional aspects of pathogenesis, in *Lyssaviruses,* Rupprecht, C.E., Dietzschold, B., and Koprowski, H., Eds., Springer, Berlin, *Curr. Topics Microbiol. Immunol.,* 187, 43, 1994.

189. **Francki, R.I.B.,** Plant rhabdoviruses, *Adv. Virus Res.,* 18, 257, 1973.

190. **Wunner, W.H., Calisher, C.H., Dietzgen, R.G., Jackson, A.O., Kitajima, E.W., Lafon, M., Leong, J.C., Nichol, S., Peters, D., Smith, J.S., and Walker, P.J.,** *Rhabdoviridae,* in *Virus Taxonomy, Sixth Report of the International Committee on Taxonomy of Viruses,* Murphy, F.A., Fauquet, C.M., Bishop, D.H.L., Ghabrial, S.A., Jarvis, A.W., Martelli, G.P., Mayo, M.A., and Summers, M.D., Eds., Springer, Vienna, *Arch. Virol.,* Suppl. 10, 275, 1995.

191. **Olkkonen, V.,** Structure–Function Relationship in the Nucleocapsid of the Double-Stranded RNA Phage φ6, Ph.D. thesis, Department of Genetics, University of Helsinki, 1991.

192. **Bamford, D.H. and Wickner, R.B.,** Assembly of double-stranded RNA viruses: bacteriophage φ6 and yeast virus L-A, *Semin. Virol.,* 5, 61, 1994.

193. **Bamford, D.H. and Mindich, L.,** Electron microscopy of cells infected with nonsense mutants of bacteriophage φ6, *Virology,* 107, 222, 1980.

194. **Stitt, B.L. and Mindich, L.,** Morphogenesis of bacteriophage φ6: a presumptive viral membrane precursor, *Virology,* 127, 446, 1983.

195. **Ktistakis, N.T., Kao, C.-Y., and Lang, D.,** *In vitro* assembly of the outer shell of bacteriophage φ6 nucleocapsid, *Virology,* 166, 91, 1988.

196. **Buck, K.W.,** Replication of double-stranded RNA mycoviruses, in *Viruses and Plasmids of Fungi,* Lemke, P.A., Ed., Marcel Dekker, New York, 1979, 93.

197. **Schiff, L.A. and Fields, B.N.,** Reoviruses and their replication, in *Virology,* 2nd ed., Vol. 2, Fields, B.N. and Knipe, D.M., Eds.-in-chief, Raven Press, New York, 1990, 1275.

198. **Estes, M.K.,** Rotaviruses and their replication, in *Virology,* 2nd ed., Vol. 2, Fields, B.N. and Knipe, D.M., Eds.-in-chief, Raven Press, New York, 1990, 1329.

199. **Tian, P., Ball, J.M., Zeng, C.Q.Y., and Estes, M.K.,** The rotavirus nonstructural glycoprotein NSP4 possesses membrane destabilization activity, *J. Virol.,* 70, 6973, 1996.

200. **Arnott, H.J., Smith, K.M., and Fullilove, S.L.,** Ultrastructure of cytoplasmic polyhedrosis viruses affecting the monarch butterfly, *Danaus flexiplus, J. Ultrastruct. Res.,* 24, 479, 1968.

201. **Wickner, R.B.,** Double-stranded RNA virus replication and packaging, *J. Biol. Chem.,* 268, 3797, 1993.

202. **MacBeth, K.J. and Patterson, J.L.,** The short transcript of *Leishmania* RNA virus is generated by RNA cleavage, *J. Virol.,* 69, 3458, 1995.

203. **Schuster, F.L. and Dunnebacke, T.H.,** Development and release of virus-like particles in *Naegleria gruberi* EG$_s$, *Cytobiologie,* 14, 131, 1976.

204. **Lazinski, D.W. and Taylor, J.M.,** Recent developments in hepatitis delta virus research, *Adv. Virus Res.,* 43, 187, 1994.

205. **Casey, J.L. and Gerin, J.L.,** Hepatitis D virus RNA editing: specific modification of adenosine in the antigenomic RNA, *J. Virol.,* 69, 7593, 1995.

206. **Taylor, J.M.,** Genetic organization and replication strategy of hepatitis delta virus, *Semin. Virol.,* 4, 313, 1993.

207. **Bruening, G.,** Replication of satellite RNA of tobacco ringspot virus, *Semin. Virol.,* 1, 127, 1990.

208. **Symons, R.H.,** Self-cleavage of RNA in the replication of viroids and virusoids, *Semin. Virol.,* 1, 117, 1990.

209. **Gruner, R., Fels, A., Qu, F., Zimmat, R., Steger, G., and Riesner, D.,** Interdependence of pathogenicity and replicability with potato spindle tuber viroid, *Virology,* 209, 60, 1995.

210. **Wills, P.R.,** Prions and theory in molecular biology, *Semin. Virol.,* 2, 181, 1991.

211. **Caughey, B. and Chesebro, B.,** Prion protein and the transmissible spongiform encephalopathies, *Trends Cell Biol.,* 7, 56, 1997.

INDEX

219

Note: Because of their extreme frequency of appearance, the following terms are not indexed at every occurrence in the text: Assembly, Attachment, Budding, DNA, Envelope, Nucleocapsid, Release, Replication, RNA, Transcription, Translation, and Uncoating.